CATS BEHAVING BADLY

ALSO BY CELIA HADDON

The Joy of Cats

One Hundred Secret Thoughts Cats Have About Humans

How To Read Your Cat's Mind

One Hundred Ways For A Cat To Find Its Inner Kitten

100 Ways For A Cat To Train Its Human

One Hundred Ways To A Happy Cat

CATS BEHAVING BADLY

WHY CATS DO THE FUNNY THINGS THEY DO

CELIA HADDON

1 3 5 7 9 10 8 6 4 2

First published in 2010 by Ebury Press,
an imprint of Ebury Publishing

A Random House Group Company

The Random House Group Limited Reg. No. 954009

Addresses for companies within the Random House Group can be found at www.randomhouse.co.uk

A CIP catalogue record for this book is available from the British Library

The Random House Group Limited supports The Forest Stewardship Council (FSC), the leading international forest certification organisation. All our titles that are printed on Greenpeace approved FSC certified paper carry the FSC logo. Our paper procurement policy can be found at www.rbooks.co.uk/environment

Illustrations: Jilly Wilkinson
Design and typesetting: seagulls.net

Printed and bound in the UK by CPI Mackays, Chatham ME5 8TD

ISBN: 9780091932152

To buy books by your favourite author and register for offers visit www.rbooks.co.uk

CONTENTS

PREFACE

PREFACE

Why cats? It used to be thought odd to like cats. Dogs were man's best friend but cats were merely second-rate pets. Kittens had a place in the nursery for the delight of children, but cats were more likely to be down in the kitchen with the servants or out in the stables keeping down mice.

Now cats are the most popular pet of all. Their cleanliness and the way they don't need walking make them relatively easy to keep. In a democratic age, they seem less class conscious than dogs. They don't cringe to their superiors – just walk away in disgust. We could learn from them.

Yet there's something very odd about the way we keep cats. Because dogs were man's first friend, it is as if dogs are the template into which we try to fit cats. There are very few drugs licensed for cats, for instance, so cats are still given veterinary drug treatment originally designed for dogs. Some 'experts' assure us that cats are social animals, almost as if they were dogs. Worse still, out-of-date behaviourists still talk about 'dominance' among cats as if they lived in packs.

We ordinary cat owners still expect dog-like behaviour from them. We make them live in groups of unrelated felines – a way of life that is fine for dogs but tough for cats. Some animal rescue shelters even keep their cats all together in a huge enclosure and even the good rescue shelters sometimes forget to give their nervous cats a place to hide away from the cats in the next door pen.

Many of us cat owners are upset if our cats don't want to be hugged and petted. After all, dogs put up with it, why can't cats? As for their ability to live with dogs, we will get a puppy without thinking twice about what it means for our cat.

We are also unreasonable in expecting them to use other people's gardens (not our own) as a toilet even though they may have to go through other cats' territories. Or we expect them to stay clean in the house even though there is one dirty litter tray for about four cats.

We may even make these hunting animals into indoor-only cats with no chance to hunt. We leave them with nothing much to do all day when we go to work, then complain if they are boisterous in the evening.

'CATS ARE NOT DOGS' is what I would like to shout. How long is it going to take before we give our cats their 'species-specific requirements', as the scientists put it.

Luckily, researchers are now doing the hard work of studying how cats operate in human society. This book could not have been written without the serious work of people like John Bradshaw, Rachel Casey, Sarah Heath, Sandra McCune, Jon Bowen, Peter Neville, Francesca Riccomini, Irene Rochlitz and Denis Turner. There are many others, of course, but these are the names that I have consulted most frequently in my studies.

I think we are on our way to giving our cats a better deal.

CHAPTER ONE
Understanding Your Cat

'It is a beast of prey, even a tame one, more especially the wild, it being in the opinion of many, nothing but a diminutive lion … being a crafty, subtle, watchful creature, very loving and familiar with mankind, the mortal enemy to the rat, mouse, and all sorts of birds, which it seizes on as its prey.'

William Salmon, *The English Physician*, 1693

Cats often outwit humans. They can understand human beings as perfectly as they need to. They are not much interested in us, but they focus on getting what they need. They know enough to get what they want from us. We, on the other hand, often don't really know very much about cats. We are so much under their spell that we can't see the feline nature behind their elegant form and charming gestures. We don't really understand cats.

Are cats really domesticated?

We used to think that prehistoric man domesticated animals. He went out hunting in the forests and savannahs, and when he killed a female animal he brought home the young ones as pets. Clever humans, dumb animals – to let themselves become domesticated.

But what about cats? As far as cats are concerned, that old story is definitely a myth. The occasional wild cat may have been picked up as a very young kitten and tamed, but it's not easy. Anybody who has tried to handle feral kittens after they have been weaned knows that picking up wild kittens is incredibly difficult. Besides, rearing kittens from a very early age would have been difficult for prehistoric man; Neolithic hunters had better things to do with their time than nurse kittens. We might like to think that we humans domesticated cats, but it's more likely that cats domesticated themselves.

Understanding this will help us understand our own cats. It's no good thinking that they are somehow like dogs or like cattle – and that we can treat them like servants. That won't work. If domestication means being at man's beck and call, cats just won't do it. They do their own thing.

Cats can live wild. You can see them all over the world – that is, if you look out for them. Some hang round restaurants in Mediterranean holiday resorts living off handouts from visitors, others live in farms and stables catching mice, still others hang out in big towns, coming out at dusk to check the dustbins and thrown-away portions of take-away food. Cats are wonderfully adaptive; as long as they have shelter from the weather and something to eat, whether it is mice or portions of hamburgers, they can survive. Even pet cats, if they lose their homes, can sometimes live for a long time in the wild.

CAT TIP

If you are holidaying at a place where there are many stray cats, keep an eye out for a Cat Café. These are little shelters – part of a scheme organized by the World Society for the Protection of Animals (WSPA) – where tourists can feed stray cats.

Cats have spread all round the globe, living on their own as well as living with humankind. Cats are everywhere. You can find them in the deepest part of the Australian bush, miles and miles away from human habitation. They live on small islands far out in the world's oceans, descendants of ship's cats that jumped ship years ago, when the sailors stopped to get water.

Domestic cats are an outstandingly successful species. All the big wild cats and most of the small ones face a bleak future. The relentless growth of human populations puts them under pressure. Many of the big cats now live only in small areas of the world. Yet their tiny little domestic cousins are everywhere. There are literally millions of them – about 600 million is one guess.

Cats move in and out of domestication according to circumstances. An individual pet cat that has not been neutered may be pushed out of its home, but he will survive long enough to start a colony of feral cats. Many cats that lose their homes will do their best to find a new one. They much prefer the easy life of regular meals, warm central heating and soft beds, to trying to scratch a living on the streets or hedgerows. Best of all, if they have a cat flap, they can have the best of both worlds – shelter and food and the chance to hunt.

If cats could choose, a home with regular meals is what most of them would prefer. Many cat owners didn't even plan to have a cat. A stray cat turns up in their garden or starts sitting longingly outside their back door. A meal or two later, the same cat moves in with them and makes it plain that he is there to stay.

This is the intriguing part of the cat–human relationship. Cats often seem to get their own way.

So did cats domesticate us?

From the cats' point of view, they domesticated us. They have managed to persuade humans to feed them, house them, and even give them medical services. Most of us humans don't ask anything in return. Cats don't have to learn to sit and beg, wait patiently for a human to pick up a lead and take them for a walk, go jogging with us, fetch balls in the park, guard us like German shepherds, herd sheep like collies or fetch pheasants like gundogs.

Cats just are there. They lie under radiators, watch out of the window, dig up our seed beds in the garden, and nowadays, rather than clear our houses of mice, they probably bring them in the first place. Cats lucky enough to have their own cat flaps are free to come and go as they please. Cats will even rehome themselves if they think they can do better further down the road.

Clever cats. They are now probably the most numerous pet in the world; in numbers far greater than their old rivals, dogs. How did they find their way into our houses and hearts?

Your cat's heritage

Understanding your cat's ancestry helps you understand your cat. We used to think that cats were first domesticated in ancient Egypt. Early archaeologists found cat cemeteries, statues of various Egyptian goddesses in the shape of cats, and cats appeared in Egyptian wall paintings. It looked as if the ancient Egyptians were so into cats that they were the first humans who brought cats into our homes. However, while Egyptians may have been the first to tame or value cats, cats were already living in human settlements.

Now we know where our cats came from – the Near East, the area from the eastern Mediterranean around Turkey, down into Mesopotamia. Scientists have worked carefully on cat DNA, and in particular compared the DNA between domestic cats and various wild cats. From this they know that the common ancestors of all domesticated cats were the African wild cats, *Felis silvestris Lybica*, that were living some 130,000 years ago in what is known as the Fertile Crescent. This is 'exactly the place where humanity settled down to agriculture ten to twelve thousand years ago,' says Dr Stephen J. O'Brien, Chief of the Laboratory of Genomic Diversity at the National Cancer Institute in Bethesda, Maryland, and one of the scientists studying the cat genome.

CAT FACT: The earliest pet cat was found by archaeologists in a Neolithic grave in Cyprus. The complete cat skeleton dated back to about 9,500 years ago. A small pit had been dug and the body of a cat had been placed in it, then covered up. It was just 40 cm (1.3 feet) away from a human grave, which had offerings such as flint tools and polished stone axes. It is possible that the cat was buried at the same time as the human, suggesting some sort of cat–human relationship, though it is also possible it was just a wild cat that was hunted down and killed. Cat teeth have been found in a couple of later archaeological sites, but evidence of full domestication comes from Egypt 3,500 years ago. Egyptian paintings show cats sitting under chairs, eating from bowls and sometimes wearing collars. A little later Egyptians began worshipping a cat goddess, Bastet. They sacrificed cats to her, mummifying them and burying them in cat cemeteries. An ancient Greek historian, Heroditus, reported that when a cat died in a household fire, onlookers would go into mourning.

Look at it from the cat's point of view: it wasn't until we were civilized that cats bothered to move in with us. That was around the time that humans stopped being hunter-gatherers roaming round the landscape to follow game and settled down in one place to grow crops. Cats like known territory, not the wandering life of a hunter-gatherer. So humans who were always on the move did not make good cat owners.

Once humans settled down, cats joined them. Stored grain from Neolithic farming and the accumulated rubbish of the human settlement attracted mice and rats. And the rodents attracted cats to these first human villages. Not only did cats find food there, but they also found shelter in the buildings, instead of having to search for hollow trees or rock cavities. They started living side by side with humans.

This wasn't a once-for-all moment in a particular location. The geneticists think that there were several moments and several places where cats started living near humans. The ancestral genes show several different lineages.

Two things happened. The cats that were least wild were most successful in living near human settlements. Wild animals have something called a flight zone. If a human moves into this area, the animal starts moving away. Even before they have started moving off, many wild animals stop eating and start staring to make sure they can be ready to run from humans.

When cats moved into human settlements as equals

The wild cats that domesticated themselves were the ones that were happy to mouse under the eye, so to speak, of the humans. They had smaller flight zones. The wilder ones probably went back to the desert from which they had originally come. The bolder ones stayed put, catching rodents among the human settlements. They reproduced and their kittens were bolder too. Little by little a whole population of tamer cats emerged. There may even have been a sudden genetic change that accounted for their tameness.

At the same time the humans must have noticed these small furry animals hanging around. Cats were helping them keep down the mouse population and a smaller mouse population meant that more grain survived in storage. It was a partnership that made sense. Cats had the benefit of some shelter and the food they preferred, mice. Humans had a natural rodent control operative on the premises. What is more, there was a sort of equality about it. Both species benefitted. Cats didn't look up to men as their benefactors then, and they still don't! Cats don't do gratitude.

This natural partnership, which wasn't yet full domestication, still exists today among farm cats. A few cats around farm buildings help keep down rodents. Even if the farmer doesn't want a pet, he may think it worthwhile to have a farm cat which lives in his farm buildings. Today's feral farm cats are living much the same way as Neolithic cats did 10,000 years ago.

When the original dogs set up home with humans, they had the group mentality. They were looking, so to speak, for a group to fit into. Dogs today aren't wolves but they look to humans for care and for instruction. They are happy to be led.

Cats aren't. Your cat does not consider you his leader. Not ever.

It's each cat for himself; the genes of a singleton

Your cat has the instincts of a separate but equal individual. This individuality is his inheritance.

Their ancestor, the African wild cat, lived in the desert areas where it had to hunt small mammals, birds and reptiles in order to eat. African wild cats don't hunt in packs; they don't help each other pull down prey because their prey is small, rather than large. Each cat goes out separately to hunt for his dinner. He will often bring back food to a safe place to eat it (because he doesn't want to get jumped on by a larger predator while he is busy eating), but he won't bring back food to share with other cats.

Wild cats, and their domestic cousins, usually don't share. The only time that *Felis silvestris lybica* shares is when a mother cat brings back prey to feed her kittens. All mammals have to share food with their young, otherwise none of their babies could ever survive.

A cat hunts for himself. He doesn't need help in getting his food. So your cat has the instincts of a solitary, not a group, hunter. It's each cat for himself. We humans call it selfishness because we think like a group animal. Cats don't have the group gene, they have the self gene. It's not selfishness to them; it's just proper self-care.

CAT FACT: Not much is known about the African wild cat, the ancestor of our domestic moggies. They are also rarely found in zoos and only one UK zoo has a pair of these interesting animals – Howletts Wild Animal Park, near Canterbury – which were sent as a gift from South Africa in 2001. In appearance, Otavi, the male, looks just like a very large, tabby domestic cat.

Within their enclosure the two cats behave much like domestic cats. The male, Otavi, sprays at various territorial marking spots and both cats scratch the wooden tree trunks to condition their claws. There is a latrine area in one corner of the enclosure. On fine days the pair spend time on a high sitting place, grooming each other and sleeping.

So far they have not bred and the zoo doesn't yet know why. 'Oranje, the female, comes on heat and there's a lot of mating behaviour but no kittens,' says Jim Vaissie, head of the cats' section at the park.

The cats are fed dead small animals, given to them whole, such as rats, rabbits, mice, chicks and pigeons. In the wild they would also eat other small rodents, insects, lizards and other small reptiles. Because the cats were sent as a gift at the time of the death of the zoo's founder, John Aspinall, little is known about their history.

Even in captivity, the behaviour of African wild cats is distinctly different from that of the Scottish wild cat, the UK's only wild feline. 'These are much tamer than the Scottish cats,' says Jim Vaissie, who has experience of both. 'The female can be quite aggressive but the male is more placid. I get the impression that they would tame quite easily.'

Sex, hunting and safety – the hard-wiring of cats

Cats wouldn't survive unless they had instincts. These are patterns of behaviour that are hard-wired into them. Instincts help animals (and humans, too) behave in ways that keep them alive and enable them to reproduce. A stud cat has to be able to reproduce otherwise his breeder will not want him on the premises. When a queen (a female cat) is brought to his cat chalet, he has to be able to perform. Viagra isn't sold for cats, because most tomcats don't need it. Instinct tells them what to do and instinct makes both tom and queen get on with it!

Of course, most of the feral cats that live all over the world simply wouldn't survive without being able to catch mice and birds. Even the cats that live out of dustbins and trash cans have instincts that will make them pounce on the nearby rats and mice that also live among the rubbish.

But as well as being predators, cats are prey to bigger animals. When a large bull terrier comes round the corner of the street, your cat needs to run up a tree out of his way. He may not even survive if he sits there dithering about what to do. It's his instinctive reaction that will save his life. And he needs to know his territory so as to get to the right tree in time.

These instincts cannot be entirely wiped out. Though the pampered pedigree cat may be neutered and live in a flat, it still has these instincts, and some of them will emerge if conditions are right. They are:

- The instinct for sex, reproduction and maternal care.
- The instinct for hunting.
- The instinct for keeping out of harm's way.

It's when these instincts come into collision with domestic life that we humans get upset. The cat is just doing what comes naturally. He is merely fulfilling his proper feline destiny. The cat behaving badly – in our eyes – is usually just behaving naturally. We need to understand these instincts in order to cope with the moments when feline instincts clash with human preferences.

Sex and the single cat

We humans can get the cat's sexual instincts under our control by neutering and spaying. Most of the time this is the best thing that ever happened, not just for cat lovers but for the cats themselves.

Before neutering was readily available for female cats, cats had kittens almost all the time. Cat lovers complained that they just couldn't keep up with the production line of these little furry delights. A single female cat can produce 200 kittens in her lifetime and, if all these and their descendants survived, there would be as many as 65,536 extra cats in the world five years later. The cat's reproductive power is awesome!

So most people, after they had given away a kitten or two, or three or four, to friends, ran out of friends who would take them. They would have to have the kittens disposed of somehow. Nowadays, thanks to neutering and spaying, we don't have to slaughter these little innocents. If anybody wants a kitten there are plenty of them available at a rescue shelter.

The other reason for neutering is to stop sexual behaviour. Female cats that come on heat are said to be calling. They become very affectionate, rubbing their bodies against you and the furniture. Then they will crouch down with their rear ends in the air, their tail raised and held to one side making it clear what they have in mind! Meanwhile their rear legs are rhythmically treading. There is also the call, a yowl – the equivalent of 'Come on in, boys. I am ready for it.' And finally, female cats on heat may squirt urine, backing up to a vertical surface, quivering their tail and letting fly.

Female cats, when they want it, usually find a way to get it. Love laughs at locksmiths in the feline, as well as the human world. These queens are accomplished flirts that will mate with a wide variety of tomcats. They are also escape artists from the tip of their whiskers to the end of their sideways-held tail. It is not for nothing that brothels are sometimes called cat houses.

Living with this behaviour every two to three weeks for several months a year and ending up knee deep in kittens, is more than most of us humans can take. So most of us spay our cats. The cats themselves become exhausted with bearing so many kittens and, if they are mating with the neighbourhood stray toms, may catch various diseases from them. Toms hold down females by biting the back of their necks and can pass on infection through the bite.

A tomcat that is not neutered is just as difficult. Tomcats roam around, sometimes for miles, looking for females on heat. They get into fights with other toms. They hang about outside any houses that contain females or sit on the rooftops caterwauling. They will spray urine where they want to leave their mark – outside the neighbour's house, inside their own house, just anywhere they feel like it.

Thoughtless humans who don't get their tom kittens neutered, often throw them out of the house when they reach the age of sexual maturity. Tomcats on the street don't last very long, they get feline diseases from fighting each other and they roam for miles, putting themselves at risk on the roads. But once they have been neutered they can look forward to a long and happy life as a much-loved pet.

So neutering and spaying will not only make a cat into a loving pet for us, but it will also extend a cat's life by several years. Neutering benefits cats, as well as their companion humans.

CAT TALE: Caesar was a big entire tomcat, with a large head and the thin, slinky body of an all-male feline. His face was covered in scars and his ears were tattered from fights with other males, but he was friendly not feral. Obviously he had been brought up as a pet when he was a kitten and had probably been thrown out when he began to mark his territory with urine, as tomcats do.

Caesar stank strongly of cat urine – a smell so powerful that it must have advertised his whereabouts and attracted calling female cats from miles away. This strong pheromone is not a smell that attracts humans, however, which is probably why Caesar lost his original home. Cat lovers, who started feeding him, couldn't take him on as a pet so they borrowed a trap from their nearby Blue Cross, trapped him and brought him into the rescue shelter.

Once in the cat chalet Caesar sprayed his new territory extensively and the smell of it wafted down the passageway between the cat pens! Luckily for him, when he was tested for FIV and the feline leukaemia virus he tested clear, so he was neutered and put up for adoption. It took about a month for the strong smell of his masculinity to fade, as his male hormones slowly ceased after neutering. He also began to lose some of his slinky tom appearance.

His scars healed and faded a little, his rear quarters began to fill out and his behaviour became gentler. As a Blue Cross volunteer, I visited him once a week and saw him change from a caterwauling king of the rooftops to a loving, calm pet. He found a home and is now living out his life in comfort, no longer in danger of disease or traffic accidents.

The maternal instinct and the inner kitten in every cat

Even though most of us will never breed cats, it is worth knowing a little about how cats bring up their kittens. If a kitten isn't brought up properly with a mother and litter-mates, he may find it difficult to behave as a cat should. But if he is brought up without any human contact, he will not be able to make a calm and loving pet. Dysfunctional kittenhoods make for dysfunctional pets.

What a kitten learns in the first eight weeks of his life stays with him forever. The cat's mothering instinct is innate, hard-wired in order to enable the species to survive. Mother cats suckle their kittens for five to six weeks. It's a joy to see the babies suckling their mother, often treading with their paws against her side and purring. Kittens need to be gently handled daily during this period, to prepare them for life as a pet. Without this handling, they will grow up wild, not tame.

When the kittens are about three weeks old, the mother cat (if living in the wild) will start bringing them food; first dead prey then living prey. In this way she is teaching them to find their own food. This is the time that cats always share their food with others. There are sometimes stories about cats bringing home food for other adult cats, but these are rare.

Weaning begins when the kittens' teeth begin to erupt, at round about the fourth week. At this point the mother may push them away from her, and the kitten learns that it cannot always get what it wants. This isn't just a lesson in growing up; it is a lesson in how to tolerate frustration. Cats that have been bottle-fed by humans often grow up to be spoiled and aggressive if their wishes are flouted (see pp. 62–3).

The hunting instinct makes every cat into a miniature tiger

Your cat, and every cat, has the mind of a predator. A cat is designed as a killing machine. The soft paws that help him move silently through the undergrowth conceal vicious claws that spring out to grab the prey. The sharp teeth can sheer through the backbone of a mouse with a killing bite. A cat kills to live, because nature has given him a digestive system that can't cope with a vegetarian diet. The cat is an 'obligate carnivore' – which means he has to eat meat.

Nature has hard-wired the feline brain with an instinct to hunt, and this instinct operates independently of hunger. We feed our cats lavishly so that they don't need to hunt for food, but their instincts don't know this. Even when a cat is full up with expensive cat food, he will pounce should an unwary mouse come sauntering by! They can't help doing it. Our dearest pussycat is still a killer at heart and we can never change that. Indeed, a well-fed cat is probably a more efficient killer, with more energy to spare for hunting.

We say that cats hunt for fun, as if there was something immoral about that. It's not a moral choice for them. Hunting makes a cat happy. All the happiness centres in his brain are fired up when he does what he was born to do – hunt. Though catching prey is satisfying, cats will hunt when the prey is just a toy being enticingly moved around by a human owner. An indoor cat that has never even seen a mouse will hunt a moving piece of string. The movement sets off his instinct and he has to hunt.

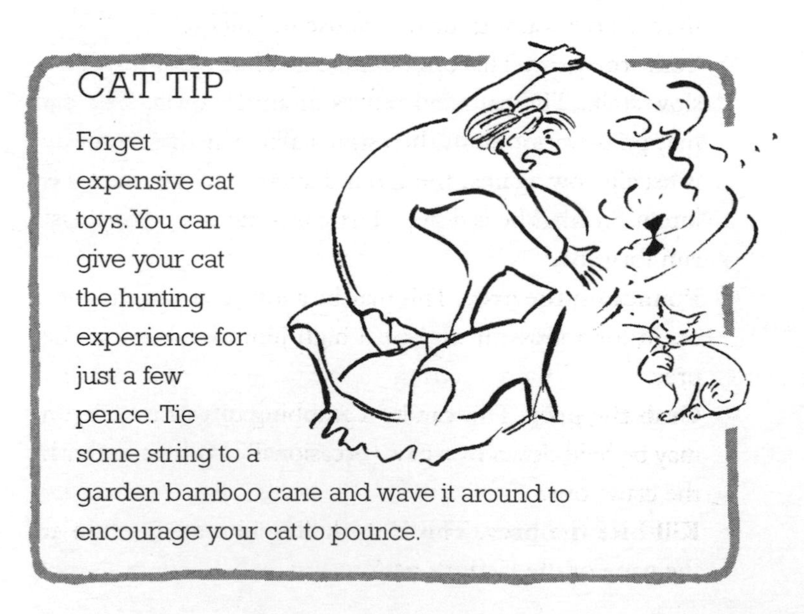

CAT TIP

Forget expensive cat toys. You can give your cat the hunting experience for just a few pence. Tie some string to a garden bamboo cane and wave it around to encourage your cat to pounce.

We also need to understand how cats hunt if we are to understand the leopard on our armchair! They don't chase and pull down their prey like a wolf. While cats can move fast in an emergency, they are sprinters, not long-distance runners. So they spend their energies, not in chasing, but in searching out and then ambushing their prey. A cat has innate and endlessly persistent patience as he waits at the mouse hole.

There is a hunting programme, called the predatory sequence, which is installed in the brain of every cat. It is a set of moves that usually occurs in a predictable serial order, with one action succeeding the previous one. It goes like this:

- **Eye the prey**. This is what a cat is doing as he sits by a mouse hole waiting for the mouse to emerge.
- **Stalk the prey.** This can be done in a flash, or it can be a slow stalk. The cat's tail moves in anticipation. The cat may pause motionless, then start stalking again. His body is usually low against the ground and the whole animal is intent on what he is doing. There is sometimes a final fast run forward.
- **Pounce on the prey**. This may be a low jump on the prey or, in long grass, it may be a high jump landing on the prey.
- **Grab the prey**. This can be a grabbing bite or the victim may be held down by a paw (occasionally both paws) with the claws out.
- **Kill-bite the prey**. This is the killing bite made often at the nape of the victim's neck.

- **Eat the prey.** This involves tearing off feathers and skin and crunching through bones and muscles. Slurping his way through tinned cat food is not the same as tearing off skin and feathers.

This sequence of actions is what a cat is all about. A wild cat hunts to live, but, because of the predatory instinct, even a tame cat lives to hunt. All these actions are immensely rewarding for the cat that performs them, even if there isn't a delicious mouse meal at the end.

CAT TALE: Feely Felix is a blind hunter and his behaviour shows how important the hunting instinct is to cats. At the age of six years old he came into the care of the Cats Protection League in Wrexham. He was completely blind, probably as a result of cat 'flu when he was a kitten, but very cuddly.

He found a loving home with Janice. 'He welcomed us at the shelter like long-lost owners,' she recalled. 'I expected some wariness at least, but not a bit of it! He wanted to rub noses and be cuddled immediately.'

Felix has settled down as an indoor cat in a flat and he spends a lot of time on the windowsill in the sun, listening to the birds on the gutters – his equivalent of eyeing his prey. Obviously, he cannot get at them through the glass but he still enjoys the first stage of the hunting sequence even if he only hears rather than sees his prey.

Indeed, even though he can't see anything at all, Felix can and does hunt, when he gets the chance. One summer he caught two flies. 'We don't get many of them in a second-floor flat, but he homed in on them immediately, negotiating the furniture without hesitation. Before I could stop him he caught them against the window and ate each individually.'

He also enjoys hunting games with cat toys, but these have to make a noise. Anything that rustles interests him and he will play with newspapers. 'He likes string with toffee papers tied to it so that it sounds like a bird when waved in the air,' says Janice. Felix doesn't need to hunt for his dinner (Janice supplies him with an excellent diet), and although he couldn't really catch mice because he is blind, he still enjoys performing his instinctive hunting moves.

Honouring the hunting instinct

You can test for yourself how happy hunting makes a cat, by moving a piece of string in front of him. Relaxed and healthy cats will immediately look alert and intent, and, if you are moving the string in an exciting way, most of them will pounce on it. This is the play hunting that is so important for indoor cats – the only chance they get to fulfil their natural hunting instinct.

If your cat catches living prey, you can also see how this hunting instinct is tied up with a sequence of actions. Each action follows the one before it and sometimes one move can't be performed until the previous moves have preceded it. Scientists call this a motor pattern, a pattern of actions or behaviours that are connected to one another.

If you have ever watched a cat making an unsuccessful pounce, you will see that he rarely pounces again immediately. Usually he will go back to the beginning of the predatory sequence of eye, stalk and pounce. He will eye his prey, do a shortened version of the stalk, and *then* pounce for the second time. The sequence is hard-wired in, so the cat normally has to follow it in more or less the right order.

Why do cats play with and torture mice?

Cats often play with their prey before killing it. Some scientists say that playing with prey has a purpose; the cat needs to get the prey into the right position for the kill bite, which usually severs the spinal chord. This final bite keeps the cat's head out of range of the mouse's teeth.

Yet performing the kill bite involves letting go of the prey. This is the moment when the prey may escape, and about two out of three mice do. Scientist and author Roger Tabor has examined the way in which a cat will play with a field vole, prodding the vole while being careful to avoid being bitten by the tiny animal. He calls it 'dazing' the prey.

What looks like torture is in fact a series of moves to manoeuvre the prey into the right position for the kill. Trying to kill a mouse by biting its backside, for instance, won't work, and the mouse is likely to either wriggle away or turn round and bite back. In the wild a hunting animal relies on being completely fit; if he is wounded or not feeling well he will not be fast enough to catch his next meal. A wounded hunter starves. Even a small mouse bite can result in an abscess or an infection. Worse still, a rat or a poisonous snake (yes, cats kill snakes) can inflict serious damage on a cat.

Playing with or dazing the prey tires out the mouse sufficiently that the cat can go in for the kill with no danger of being bitten. And during this play we see the animal going through the first few moves of the predatory sequence of eye, stalk and pounce over and over again.

Can we stop a cat being a hunter? I don't believe we can. We can take the cat out of its hunting ground but we can't take the hunting instinct out of the cat. Keeping a cat indoors all its life will stop him killing mice and birds, but even so, many indoor cats will hunt flies on the windowsill and some very bored cats may start hunting their human owners, lying in wait for them and ambushing them (see pp. 67–8, 96–109). Cats need to hunt, and, if they have no chance to do so, some will start behaving badly in painful ways towards their humans. A wise owner will help cats fulfil this need, even if the prey is just a bunch of feathers on a stick.

Understanding territory – the need for a safe home

Cats are prey as well as predators, hunted as well as hunters. Keeping safe is the third hard-wired instinct. While lions may be king of the jungle, our cats certainly are not. They are small hunters who are hunted by larger animals. The original *Felis silvestris lybica* makes a nice meal for a bigger predator. Among its enemies are hyenas, jackals, wild dogs, lions, birds of prey and large snakes. So while the domestic cat has inherited the hunting instinct from his ancestor, he has also inherited the caution of the hunted. The predator, much like the mice he hunts, must avoid becoming prey.

When danger threatens, cats have four options – fight, flight, freeze, or fiddle about (i.e., do something submissive or odd which might make an enemy decide not to kill you). Fighting isn't much of an option for a cat. Even with something about their own size, like a terrier or a fox, they are likely to be severely injured. Nor, being small, can they bluff it out by a threatening show of force – though they will fight if all else fails.

Cats usually run away, often up a tree or to a high place. They feel safer high up. They also avoid danger by keeping their distance from it; by hiding, by searching out a safe familiar territory, and by knowing their surroundings perfectly. Knowing where the nearest high tree is may save their lives. Cats will occasionally stand their ground, but this usually occurs when they have nowhere to run. Their sharp teeth and sharper claws are no defence against enemies many times larger than themselves.

To feel safe, cats must have a familiar home territory – a place to sleep in safety, a place to bring back prey to eat in safety, a place to have their kittens, and a place just to hang out. To feel safe in a house, cats need hiding or sleeping places, spots where they can leave their own scent by rubbing, and scratching areas where they can leave scent messages for themselves or others. A house is not a home unless it smells familiar to the cat. The cats that live with you and me leave their scent on our furniture, on our bed, and on our ankles and laps. They mark us, as part of the family, with a family smell.

To a cat, familiar is reassuring, but new is frightening. Cats hate the shock of the new. That is one reason why, when we move house, we must keep cats inside a new dwelling for two to four weeks. It won't become a home to the cat until he has explored every nook and cranny, and left his scent in all the

CAT TIP

Give your cat time to re-examine his territory when he comes home from a visit to the vet or a cattery. He will walk round the house and garden, carefully inspecting various key territory marks and boundaries. He may then re-mark the various territorial marks by rubbing, scratching or spraying, to show he is back and to make sure his territory has his own smell.

correct (to a cat) places. Let him out too early and he may just run off, because he doesn't feel the new house is a safe place to come home to.

As well as safe places to sleep and eat, cats need safe places to urinate and defecate (see pp. 200–23). After all, a cat that is squatting is very vulnerable to being ambushed. All cats should have a litter tray inside the house, even if they mostly go outside for their toilet. Young cats, elderly cats, or cats that have been temporarily frightened may need to use a litter tray, where they can feel extra safe.

Understanding territory – the hunting range

Sleeping, eating and hanging about are all done safely at home. Hunting is done on the hunting range. This too has to be safe, but it doesn't have to be quite as safe as the home territory. Sometimes several cats that don't live together will share a hunting range. They will not interfere with each other and will all keep a prudent distance apart. The hunting range isn't a unitary flat piece of territory like a field; it's a series of pathways leading to patches of hunting territory. Cats may even work out a sort of time-share so that they visit the same places at different times.

The hunting range may be your garden or it may be further away, perhaps even in the wood or the field across the road. Some younger, bolder cats will range far and wide to hunt, up and down hedges, in country areas several fields away from the house. The bolder they are and the further they go, the greater the danger. Many cat owners don't even know where their cat goes while he is out hunting.

Accept, don't try to change, the feline instincts

The three essential feline instincts for sex, reproduction and maternal care, hunting, and keeping out of harm's way are what make up a cat – that beautiful elegant tiger in your house. If you ignore, thwart or frustrate your cat's essential nature, you will end up with an unhappy stressed-out cat. If you work with his instincts, your cat will be happy and relaxed.

CHAPTER TWO

Cats Versus Humans

'The cat is domestic only as far as it suits its own ends; it will not be kennelled or harnessed nor suffer any dictation as to its goings out or comings in. Long contact with the human race has developed in it the art of diplomacy, and no Roman Cardinal of medieval days knew better how to ingratiate himself with his surroundings than a cat with a saucer of cream on its mental horizon. But the social smoothness, the purring innocence, the softness of the velvet paw may be laid aside at a moment's notice, and the sinuous feline may disappear, in deliberate aloofness, to a world of roofs and chimney-stacks, where the human element is distanced and disregarded.'

Saki (*Hector Hugh Munro*), 1870–1916

All cats are wildly individual. Indeed, some are downright eccentric. Anybody who has kept a succession of cats knows how strong and how different is the personality of each. The closer and longer your relationship, the more likely it is that your cat has developed some very interesting and funny ways of behaving. Cats blossom into truly amazing characters if they are encouraged to do so. For full expression of a cat's individuality, she needs a home which gives her emotional serenity and lots of choice.

Cats love choice. You can give a dog a new bed and tell her to go and sit there and she will do so without complaint. Buy a cat a new and expensive bed and she will refuse to use it for months and months. You can put cat treats in this new cat bed and she will eat them. Then she will swiftly retreat as soon as there are none left. Most domestic cats choose where they will sleep – on clean linen in the warmth of the airing cupboard, on windowsills where they can see the local pigeons flying about, high up on the top of cupboards where they cannot be seen. Some of the more eccentric sleeping places chosen by cats include wastepaper baskets, between the computer keyboard and the screen, inside open box files, on the top of the back of sofas, inside drawers, or snuggled down *inside* the bed.

CAT TIP

Cats will sleep in extraordinary places – wastepaper baskets, cardboard boxes, baskets and other odd containers. Keep an eye out for these at boot sales. If you buy an expensive bed, your cat probably won't use it!

It is the cat's love of choice that explains why they dither in doorways. Even cats who have cat flaps in their homes often prefer to have a door opened for them. They will sit patiently waiting to be let in or out. But once the door is opened, they pause, half-in and half-out, as if they are now undecided whether to come in or go out. Sometimes, to the acute irritation of us humans standing at a windy open door, they appear to change their minds, deciding that they don't wish to come in (or go out) after all. They just wanted to know that it was possible to do so. This isn't as crazy as it seems to us humans.

Human etiquette says it is impolite to linger at a door being held open; feline etiquette is different. For a small animal, security is important and one way to ensure this is repeatedly to check the territory, making sure that entry or exit into the safe home area is possible at all times. A cat behaving badly (to us) is usually a cat behaving perfectly properly (by her own standards).

CAT TALE: Charlie is a handsome black-and-white cat who has set up several homes. He can often be seen hunting in the fields or sitting in the back gardens of a row of houses. He has three owners. His real owner is a woman who worked in the nearby council offices, who picked him up as a stray, but two other people within the village also think he is their cat. He comes home at night to his official owner, but spends his day with two other easy-going households who think he is out hunting at night. None of the three 'owners' know that he is 'owned' elsewhere. I discovered Charlie's two-timing (or three-timing) when I was checking every house to see if I could find my lost cat. I have not told any of the 'owners' what Charlie has been doing – they will be happier if they believe they are the only humans in his life!

Give and take; you give, the cat takes!

Of course, most cats are dependent on human beings for much of what they want. They rely on us to feed them, put down clean water, supply clean litter trays, give shelter from the cold, keep them safe from dangerous dogs, and allow them the space they need to avoid cats they do not like. If these conditions are not available, they will adapt themselves as best they can. If they don't like the food, they will raid the dinner bowls in other cats' houses. If the litter tray is wrong for them, they will use the space behind the sofa or go behind the door.

Cats will even set up two households if necessary. Many cats, whose owners go out to work and turn the central heating off, go down the road to find a house where the central heating is on during the day. They start spending nights with their official owners and daytime with their unofficial owners. Cats quickly learn how to time-share households.

If they don't feel safe at home, cats will rehome themselves. If a dog suddenly joins the household and starts chasing them, they just move. Or, if the household gets too full of cats (and for some individuals more than one cat is one too many cats), they will search for a home where they can be the only cat.

If we want our cats to behave well, we must give them the right conditions for a fulfilling life. We can't put a cat into the wrong environment and expect her to be happy. This is the human side of the bargain. If things go wrong, it is up to us (not the cat) to change. If we change the situation, usually the cat will stop behaving badly.

Stopping cats behaving badly has to begin with us. We have to understand them and then change what we do, so as to make them happy. Only if we make changes will they change.

CAT TALE: Henry, a white-and-black stray cat, was a supreme survivor who turned up in my garden a few years ago. He was a large cat with a stumpy tail – he had probably lost two-thirds of it in a traffic accident. His method of making friends was to roll onto his back as soon as he noticed you looking at him. Unlike some stray cats, he was relaxed and loving. He liked being picked up. Somewhere in his early life he had been loved and handled as a kitten.

Inquiries at the houses two fields away revealed that Henry had been living off food put down for two rottweilers. If the dogs didn't finish their meals, which were fed to them in the garden, he would hop down from the fence and eat. It was a dangerous life. The dogs tried to attack him when he was on the fence, and if they had caught him that would have been the end.

Henry had slept overnight in some of the other houses that had resident cats and cat flaps. He would wait until the family went to bed, then pop in through the flap. In the morning they would find him happily settled on the sofa. Occasionally, he used their houses in the afternoon and one friend of mine met him coming up the stairs with the evident intention of an afternoon nap on her bed.

For a time he stayed with me, living in my garden shed until I could find him some new owners. My then elderly cat, Fat Mog, was deaf and suffering from arthritis. She had never been keen on other cats and so I couldn't let Henry take up residence with me.

I found some new owners for Henry but he didn't take to their lifestyle. He took off on his travels again and went over three fields in the opposite direction to a different village. There he found a home to his liking. He is much loved and has never found it necessary to stray again – though occasionally he nips into other people's houses to steal a little cat food.

I have always admired Henry's can-do attitude. Not only did he survive the rottweilers, but, when a new home wasn't exactly what he wanted, he left in search of a better one. A good example of a cat making his own choices.

Cats are not dogs

For thousands of years cats were mainly valued as pest controllers rather than pets. They lived outside much of the time, coming to the back door for a supplementary dish of milk or scraps. In the big houses of the British upper classes, the family dogs were upstairs with their owners while the cat was kept in the kitchen and fed by the cook. Dogs were best friends and upper-class while cats were servants and lower-class. For hundreds of years dogs were the primary pets.

So it is not surprising that we are still unconsciously or even consciously influenced by the dog–human relationship when we think of cats. The veterinary world for years existed mainly for dogs; until recently most veterinary treatments for cats were dog treatments adapted or merely tried out on cats. Even animal-behaviour experts thought of cats as slightly different (or even defective) dogs and applied the ideas of pack theory and dominance (now also thought to be quite wrong for dogs) to cats. This is slowly changing.

Cats are now the primary pets in countries like the United Kingdom. If we want them to be happy, we can't treat them like dogs. They are not dogs. They do not respond like dogs. Dogs adapt themselves to human beings quite well – they want to please humans. Cats don't really see the point of either pleasing or displeasing humans.

CAT TIP

To tell how cats feel about each other, watch how much space they keep between themselves. Cats that dislike each other will keep a good distance between themselves, refuse to pass each other in doorways, and eat at separate times at the food bowl.

Cats don't do packs

The most common misunderstanding between humans and cats is about social life. We used to think of cats as solitary beings, like *The Cat That Walked By Himself*, a character in a Rudyard Kipling story. Then some behaviourists started believing that cats were social animals like dogs and they lived in groups and had dominant and submissive relationships. Neither extreme is true.

Cats are flexible and individual in their living arrangements. If there is only enough food for one cat, a feral cat will live a relatively solitary life, except for at mating times. But if feral cats have a source of food – rubbish dumps, dustbins, or a friendly feeder – a group of them will come to feed. You might, for instance, find a number of feral cats all living in a barn if the farmer puts down some food for them. They are often a family group of mothers, sisters and kittens with a visiting or temporary tomcat.

If you studied them carefully, you would usually see that they were definitely not closely bonded like a pack of wolves. They don't need to be because they hunt individually, not together. Scientists have spent a lot of time investigating whether there is a top cat – like a pack leader – and the results have often seemed to differ from one group to another or even from one time to another. There is no strict hierarchy.

Within a larger colony, such as cats living in a dockyard, the cats usually organize themselves into female family groups, with males moving between the groups. And, as well as family ties, there may be long-standing friendships between some individuals. So what looks like a large group of cats is often a collection of families of cats that share the space with others because that is the place where there is enough food or shelter for them all. Some of them are living separate lives in the same space! All of them, however, are wary of cats that are unfamiliar. Strangers are not welcomed into a colony, though if they persist they may finally be accepted.

Yet we expect cats to live with cats that are not family. Many of our cats, which are forced to live together in a house, organize their lives to avoid each other – preferring separate beds, eating at a time that other cats are not there, and using litter trays in a time-share way. They live together but with a minimum of socializing. When we get a new cat, we expect our existing cats to accept the addition of strangers into their territory – not a natural reaction at all for them.

That said, close family relationships or friendships do exist between individual cats. Some of the cats from the same household will share the same bed and groom each other when in the rescue shelter – a sign that you are adopting a pair of friends. In feral cat colonies, sister cats in a colony may sleep together or even help raise each others' kittens. But these friendly family relationships differ between individuals and there are no overall rules of rank and status that apply to all cat groups. Cat groups and families, like individual cats themselves, vary enormously.

Some cats are sociable. Some aren't. While some cats will live happily in a group, others are true loners.

CAT TALE: Jaffa was living in a semi-detached house in a posh Surrey suburb where there were three other cats. The neighbours, who had no cat, began to find Jaffa in their garden. A handsome ginger neutered tom, he seemed to want to come into their kitchen, so they began letting him in. Sometimes they fed him and he often had a nap on their bed in the afternoon. At night, because they knew he lived next door, they would shut him out, believing that he went home to sleep. After several months, they went away on holiday and on their return they were horrified to find Jaffa, thin and bedraggled, waiting for them to come home. He entered their house with visible signs of relief, ate a huge meal, and collapsed on their bed.

After talking to the people next door, Jaffa's chosen new owners were given permission to take him over. It turned out that, far from going back home at night, he was simply finding shelter in the garden shed. He had, for all intents and purposes, left home where he was being bullied by the other three cats. Rather than put up with this, he had decided to rehome himself. Now he lives very happily indeed as a single and much-loved cat.

Cats look down on us

The old joke goes that dogs look up to us but cats look down on us. Certainly, my own cats have often given me the impression that they see me as a rather tiresome social inferior. No cat I have ever owned has made me feel that I was superior.

As our cats don't look up to leaders or alpha cats, they don't look up to us, either. Dogs know how to be submissive and crawl on their bellies in front of superiors, but cats don't. Their method of dealing with trouble is to avoid it in the first place, but, if a quarrel breaks out, they have no submissive body language to say that they are sorry. Dogs can usually quarrel and make up. Cats often can't. They have long memories and some cats do not do forgiveness – ever.

This is also why punishing cats is a complete waste of time and effort. Punishment, as any good dog trainer will tell you, is the least effective way to train dogs, let alone cats. However, dogs will put up with being punished and will even lick the hand of their abuser. Cats do not take punishment if they can avoid it. Unless they are indoor cats that cannot escape, they move out of homes when they are ill-treated.

Cats that have been hurt may never trust their abuser ever again. And some cats lose their trust in all human beings.

Do cats train their humans?

Very few humans train their cats: almost every cat trains his human. Take cat food: there is a huge variety of cat food on the market, ranging from cheap to expensive, from dry to wet, from fish-flavoured to meat-flavoured. The cat can't go shopping and buy her own food; she has to train her human to buy the correct brand. Most cats do this almost effortlessly. Ask anybody what they feed their cat and most people will

CAT TIP

A quick way of guessing whether a stray cat has an owner is to feed her some cheap cat food. If she eats this greedily, she may be a true stray. If she eats only a little of it, she has obviously had a better meal elsewhere.

answer with one or two brands of cat food. Ask them why they buy that brand and usually it will be because the cat prefers it. Naturally this is often the most expensive brand!

The feline secret weapon – persistence

The cat's most effective training technique is single-mindedness. Persistence is hard-wired into their brain as a necessary hunting tool. Feral cats have to wait at mouse holes until the mouse emerges in order to eat, so waiting comes easily to them. Unless something distracts her, a cat will wait happily at a window or door for a very long time. Sometimes they are very difficult to distract from their aims. We all know that, when we pick up a cat from the kitchen surface and put her on the floor, she will usually jump back up immediately. Or, if we shut the cat out of a room, she is likely to be found just waiting outside to get back into it. Cats can usually out-wait human beings in order to get what they want.

In waking up their humans, cats show an enviable and quite striking persistence. They keep going with their chosen technique even when the human groans and rolls back under the duvet. People whose cats wake them up for a snack in the early hours of the morning could put a stop to this by simply ignoring the cat's demands for two or three weeks. But most humans do not persist long enough in ignoring the early hours' feline summons. The cat, on the other hand, will continue to pester and wake up her humans at 3 a.m. She out-persists her owners almost every time. Cats usually win.

CAT FACT: Many cats that sleep on the bed decide when their humans should wake up. This can be extremely irritating, as their idea of a proper waking time does not allow for weekends, changes in routine or clock alterations.

- Scratching the side of the bed. 'It's like the speaking clock – at the first scratch it is 7 a.m.,' says the cat lover.
- Lying under the bed and scratching upwards.
- Scratching the dressing-table mirror to make an excruciating noise.
- Banshee wailing from cats with the command tones of a Siamese.
- Purring loudly an inch away from the ear.
- Loud purring accompanied by vigorous kneading of the human body.
- Vigorous and painful licking of ears, nose or lips – whichever is within distance of the feline tongue.
- Poking with a paw, claws sheathed.

- Poking with a paw with claws unsheathed – gently, at first.
- Walking up and down the human's body.
- Jumping from the bed head on to the human body.
- Presenting the backside, in a friendly flirtatious way, about two inches away from the human nose – surprisingly effective for those with a good sense of smell!
- Sneaking up from the bottom of the bed, wriggling under the duvet there in order to bite the human's toes.
- Vigorous feline washing while sitting on the human chest.
- Standing on the bedside cupboard and swiping off whatever stands there, one item at a time.
- Lifting the human's eyelids.
- Jumping onto the full human bladder.

How cats make their wishes known

Learning to read your cat, by learning her language, will help you behave in a way that cats appreciate. Human beings talk to each other using a complicated series of sounds when they are face to face; cats use body language and scent to talk to each other. They also use sounds, but these are not really the most important method of communication for them. Apart from the sexual sounds made by mating cats – both male and female – the most common sounds include:

- **Chirruping and trilling.** Mother cats chirrup or trill to their kittens. Some cats chirrup to their humans.

- **Miaowing.** This is the sound cats use for their humans. We all recognize it but it doesn't have a single meaning. Cats rarely use it between themselves. It can be a greeting sound, an interrogatory sound, or just an attention-getting sound. A smaller, higher version of this sound is used by kittens when in distress.

  Some cats are almost completely silent, rarely, if ever, miaowing. Others, especially Siamese or breeds with Siamese ancestry, miaow almost constantly. Older cats sometimes get more vocal when they discover that humans respond to sound.
- **Growling or snarling.** This is the sound of an angry and also often frightened cat. The snarl is higher in pitch than the growl, but its meaning is the same.
- **Yowling.** A low, loud noise, like a long-lasting and ululating miaow, from an angry cat. One version of this sounds more like a howl. Some cats also use this as a complaining noise.
- **Hissing and spitting.** These are the sounds of a cat on the defensive, warning an animal or human to back off.
- **Purring.** All cats purr, but some purr so quietly that it cannot be heard – though the vibrations can be felt if you put your ear to the cat's body. There's something magic about purring; it's a very soothing noise.

  People think that when a cat purrs, it is happy. Normally it is. Cats will purr when they are sitting on their human's lap, or when they are kneading a soft cat bed or a blanket. But they will also purr when they are not at all happy, such as when they are being handled by a vet. It is a contact noise, usually happy, originally between mother and kittens. But this message of 'I'm here, you're here' can also be used by a frightened cat when being picked up by a human.

That said, many cats don't purr at vets and some cats just bite them!

Body language – the essential feline communication

Humans are poor readers of body language as it is not our main system of communication. However, it is possible for the majority of humans to read the most obvious cat signals, and the better we know a cat, the better we are at reading feline signals. Cats, however, are experts at reading body language. They can read tiny alterations in body posture, ear position and tail posture – signals that we may find difficult to interpret.

- **Ears.** Ears held back either high or flat are a sign that the cat is feeling either aggressive or is on the defensive.
- **Tail.** When a cat is in a relaxed but active state of mind, the tail hangs low and outwards. A tail held upright, as a cat walks towards you, with the tip slightly curving forwards, is a friendly greeting signal.

  A tail tucked away between the hind legs shows that the cat is ready to defend herself. When the tail is held high initially but then tucked down, the cat is aggressive.

A swishing or lashing tail is a sign of impending aggression. If you are petting the cat and see this tail signal, stop! A sharp twitch of the tail is also a warning sign.

A quivering tail held high precedes a urine spray. There is also a dry quivering tail, usually a sign of excited affection towards a human. If the cat backs towards you with butt up high, front of the body low, and the tail held to one side, she is flirting with you. This is the come-on posture when a cat is on heat, but neutered cats sometimes use this as the language of love towards their humans.

- **Eyes.** A steady glare can be used by cats to intimidate other cats. It is polite to look away, rather than stare, in the cat world. I have seen a cat in a rescue shelter growl at passing humans who stopped to stare. It sometimes helps reassure timid cats if you don't look directly at them.

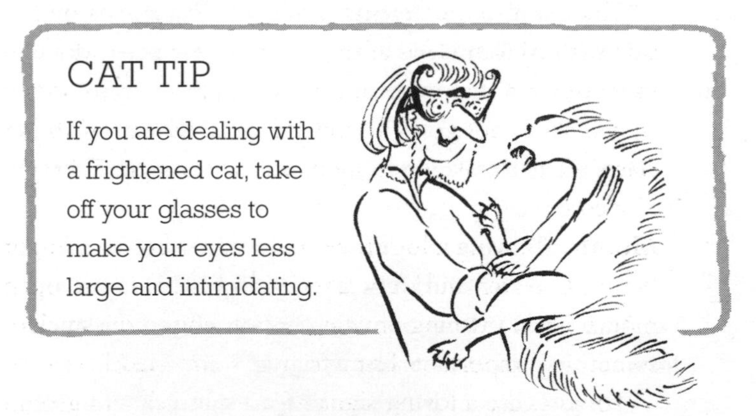

CAT TIP

If you are dealing with a frightened cat, take off your glasses to make your eyes less large and intimidating.

- **Body.** When cats fluff up their body fur and bristle the fur on their tails, they are making themselves look bigger to show that they might attack. Their backs arch, their necks stiffen, and often their mouths open to show their teeth.

Claws at the ready are also a bad sign. A raised paw, like a raised human hand, is often a threat. (However, it is a sign of affection or pleasure when cats knead with their claws. This is kitten language, used by a suckling baby cat.)

Cats that don't want to fight crouch low on the ground, flatten their ears, and make themselves look as small as possible. If they are able, they may back away carefully, still facing their opponent, to put more distance between themselves and the enemy.

Rolling on the ground can be a friendly come-on to humans or to another cat. The cat will often be purring, stretching, and looking up in a playful way. This is the social roll.

There is also the defensive roll; when a cat lies on her side with all four claws at the ready. This is often done in front of a cat that looks aggressive. The one lying down signals she is not going to initiate an attack but, with her claws ready to rake her opponent, she can defend herself if need be.

- **Mouth**. Showing a lot of teeth is a sign that these might be used! Watch out! The exception to this is the open mouth when yawning or when responding to the smell of something important, like a tomcat's urine mark.

  Tongues are a loving signal. Cats that lick and groom each other, or lick and groom their humans (it can be quite painful) are expressing their affection. Sometimes cats being stroked will put out their tongues a little and dribble, as if they were kittens.

- **Nose.** Cats will greet each other nose to nose – usually a sign of friendship or, at least, companionship. They also sniff each other's backsides, though not as intensively as dogs do. Sometimes the cat that is being sniffed in this way resents it and turns with an uplifted paw, ready to strike.

The mysterious language of scent

Although cats have a far better sense of smell than humans, we humans have only just begun to understand what scent means to a cat. There are scent glands all over the cat's body – beneath the chin, on the cheek, at the corners of the mouth, each side of the forehead, along the tail and between the toes. These give off a scent that most humans cannot smell, but which is easily perceived by another cat.

When a cat delivers a scent message by rubbing her cheek or body against her owner, most humans receive the tactile message but miss the scent message that goes with it. They simply cannot smell well enough to notice the scent. When a cat rubs against another cat, she leaves some of her scent behind. She also picks up the scent of the other cat onto her own body, mixing the two scents together. Rubbing is a sign of friendship.

This mixture of scents is the smell of family or home to a cat. It is very important. We humans recognize our friends by their faces, but scent is the way that a cat identifies friends and foes. A family member that smells wrong may be wrongly identified as an enemy. For instance, a cat that has been to the vet and smells of the hated veterinary surgery, may be attacked as a hostile intruder by a cat left at home. She may look the same, but she smells wrong.

Rubbing is also used as a territorial mark to spread the friendship message around the home. If you start watching your cat carefully you will see that she rubs against furniture and sometimes doors. Knowing where your cat rubs may be helpful if your cat gets very stressed – more on this later.

Grooming also probably plays a part as both a tactile and a scent message, as the scent from saliva would be left on the

CAT TIP

If you are bringing home a cat from a rescue centre, ask if you can take home with your new cat the material in her bed This way, her bed will smell reassuringly of home. You can return the material, washed, to the rescue centre later. Or, if there is a delay between adoption approval and taking home the cat, take her new bed to the rescue shelter so that she can get used to it there.

groomed cat's fur. Grooming between cats is a sign of friendship – enemies don't groom! So if your cat insists on grooming you, take it as a compliment – if an uncomfortable one!

Scent messages that even reach the human nose

The only scent messages that get through to the normal human nose are the strong ones – those delivered by urine or faeces left at a site close to the human. These scent messages are normal cat-to-cat communication, but we humans still don't know exactly what they mean. One in ten cats, including males, females and neutered cats, spray. Spraying allows cats to send messages without actually having to see each other. A cat may spray urine to mark the pathways in her territory and will top up the message by spraying again the next time she is there. Other passing cats will also spray on the mark, in the same way that dogs lift their legs where other dogs have done so.

Does the spray mark mean 'Keep Out'? Or does it mean, 'This is my territory'? We humans cannot be sure. It may just say, 'I was here two minutes ago so I am still around.' Or just, 'Tomcat was here.' Cats that pass by these feline messages can probably read the gender and sexual status of the cat that left the message. They can also probably read how long ago the message was posted.

Messaging without having to meet face to face is important for a cat. It allows the cat to avoid conflict by keeping a safe distance. Normally cats don't want to meet other strange cats, so these messages may help them avoid such meetings. Of course, sometimes the urine message, if it is left by a female cat on heat, actually suggests, 'Come up and see me some time.'

Spraying is also used if a cat has had an unpleasant encounter or feels her territory has been invaded. After chasing an intruding cat off his territory, a tomcat will spray. (This is probably a 'Piss off' message.) He will also spray if he can smell the previous presence of an intruder, or if the familiar scents of home have been in some way disrupted. So, if your cat starts spraying inside your house, this may be a sign that she is worried about intruders.

Finally, some smells will prompt a cat to spray. They will spray where there has been chlorine or ammonia cleaning fluids used, and they will also spray on box trees because these smell rather like cat urine (more on p. 216).

CAT FACT: Some cats sniff drugs! Catnip is the most common one. The active chemical in catnip is neptalactone, a mild hallucinogen. Cats sense this through their vomeronasal organ, an extra scent organ at the back of their nose. Sometimes they give the Flehmen response, pulling their gums back, pressing their tongue against the roof of the mouth to force the air through the vomeronasal organ – to concentrate the smell there.

Cats sniff catnip then roll around in temporary ecstasy. The enjoyment lasts about 20 minutes and then the cat is satiated. Unlike human drug addicts, cats seem to know when to stop. They only use these scents recreationally! Some cats have no interest at all in any of these scents.

Other cats are turned on by nail varnish, pears, deodorant, liquid Vapex, Deep Heat rub, Olbas oil, Domestos, Dettol, Vick vapour rub, human insulin, olives (especially stuffed with anchovies), amaryllis plants, valerian, *Actinidia kolomita*, the *kolomite* vine, *Actnidia chinensis*, Neopolitan cyclamens, bog beans (*Menyanthese trifoliata*), and (strangest of all) 'the smell of swimwear on our return from the pool'.

CAT TIP

Cats are particularly sensitive to poisons. Never use near cats phenolic compounds, such as disinfectants that turn cloudy in water or coal tar shampoos.

- **Middening (or poo messages).** One of the reasons why cats make such splendid pets is their cleanliness. Cats usually bury any faeces that they leave within their home territory. If they defecate in the hunting range, outside their home territory, they will often leave these unburied. Some scientists have suggested that unburied faeces may be a 'Keep Out' territorial mark, but this is difficult to prove.

  Our domestic cats usually use a litter tray without difficulty, if a suitable one is provided. Just occasionally there are accidents, or faeces might be found outside the box. If this happens, it is sometimes just a sign that the litter tray is not to the cat's liking. But it can also be a sign of the cat marking her territory.

We humans cannot read these scent messages correctly; we just don't understand. Instead, we get very upset, and sometimes very insulted. Yet urine and faeces messages are not meant to offend or insult.

Inside the house, spray messages are usually a cry for help from a cat that is very worried indeed. Spraying and middening are, to the cat, a solution to a problem – they reassure a worried feline – but they are a problem for humans. More of that later (see pp. 200–23).

- **Scratch and sniff messages.** Cats also mark their territory by scratching. This is a visual message but, again, it is also a scent one. Scratching trees, furniture or carpets leaves both a visual signal and also a smell, since there are scent glands between a cat's toes. So a scratch is another message board for any cats passing by.

  Cats also scratch in the presence of other cats, which may be an ostentatious display of confidence. When one cat has finished scratching, her companion may then take her place at the tree. Cats that scratch furniture in the presence of their owners have discovered that it is a way of getting human attention, if not human approval.

  Humans sometimes notice that if there have been builders in the house or unknown visitors, their cat will use her scratching post more often than usual. Or she may start scratching the furniture, the door or the mat. If strangers have been within the territory, a cat scratches to put back, or to strengthen, her own territory marks and to spread her own smell within the territory. Home has to smell right or it isn't home.

Once again, scratching is normal cat language but we humans neither understand it nor appreciate it. In fact, in countries where such operations are still legal, some humans go so far as to have the claws of their cats surgically removed to prevent them scratching.

CAT TALE: Ziggy was a cat who used his scratching post a lot. He rarely, if ever, scratched the furniture. Naturally the post began to show signs of wear and tear, with bits of rope hanging off it. His owner, Jo, had decorators in to paint the living room and decided a nice new scratching post would go with the new décor. It was a mistake. Ziggy refused to use it and started scratching the sofa and armchairs for the first time ever. I advised her that Ziggy was probably feeling rather stressed by the decorators, strangers in his house and he was scratching his territory more than ever. I suggested she put back the old scratching post, which luckily she had not thrown away. She did so and Ziggy went back to using it. In Joe's eyes the scratching post was disgracefully tatty. In Ziggy's eyes the new post wasn't good enough. It didn't smell of feline scent. The old post smelled wonderful and its tattiness made it extremely pleasurable to scratch.

The wonder of whiskers

Cats whiskers are used for receiving, not delivering, messages. The wonder of whiskers is their sensitivity. These are long strong hairs with deep roots that are connected to nerves which transmit information to the feline brain. There are whiskers on the cheeks, eyebrows, and front legs. The cheek whiskers are movable and are used when a cat catches prey; they protrude forward as she springs on a mouse and then, when the prey is seized, are used to monitor what the mouse is doing while it is dangling from the cat's mouth.

A blindfolded cat can use his whiskers to locate a mouse and grab in the right area: the nape of its neck. The moment when a whisker touches the mouse to the moment when the cat springs takes only one tenth of a second. The whiskers also help the cat in deciding which end of the mouse to eat first, i.e. from the front, so that she eats along the lie of the mouse's hair. There are also whiskers on the back of a cat's forelegs; these are presumably helpful when a cat has a mouse pinned down under her paws.

Cats also use their whiskers to feel along the body of other cats, in a friendly way, when they meet them. Yet if a cat is sleeping, touching her whiskers won't wake her, which suggests the cat can switch her whiskers on and off. Yet another amazing cat ability!

CHAPTER THREE

The Educated Kitten

'A house is never perfectly furnished for enjoyment, unless there is a child in it rising three years old, and a kitten rising six weeks...A kitten is in the animal world what the rosebud is in the garden; the one the most beautiful of all young creatures, the other the loveliest of all opening flowers.'

Robert Southey, 1774–1843

From the moment you bring home a new kitten, you are teaching him how to behave, whether you know that or not. And, though you won't realize it, your kitten is teaching you how to behave too. If your cat is not to run rings round you in his later life, it's important that you get into the teacher's seat first.

Well before he even arrives at his new home, your kitten has learnt several lessons which will influence how he behaves as an adult. The single most important lesson that your pet cat needs is how to bond with human beings. If he doesn't learn this as a kitten, he may never learn it for the rest of his life. He may always be wary around human beings and will grow up as a feral cat, and most feral cats can't be handled.

There is a window of opportunity in kittenhood, known as the sensitive period, when a cat can learn that humans are friends. Tiny kittens do not know fear. You can handle them and cuddle them much more easily than older cats. They don't run – if anything they approach. Their little mews, playfulness and lovingness are nature's way to induce their mothers, or even their humans, to look after them. We all love kittens, and luckily the human desire to pick them up and cuddle them is just what they need to become good pets.

This fear-free window falls between the second and seventh week of kitten life. The process of accepting humans as friends is often called 'socialisation', and this has to take place before kittens get the fear instinct. This instinct kicks in around the eighth week, growing stronger until the fifteenth week.

So, during the five-week period before he knows fear, your kitten needs to be handled by at least four different humans – preferably including men, women and children. The more handling your kitten has had, the friendlier he will be in later life to humans. He also needs to get used to the smell of humans, since as I've explained already, smell is important for the feline identification of friend or foe.

If he is to be adopted into a home with a dog, a kitten should meet a friendly dog during this period. If he is going to be happy in a home which has lots of other cats, he needs to meet other adult cats, not just his mother, during his kittenhood; that way he will learn how to enjoy feline social life! An only kitten brought up in isolation with just his mother, may become a bit of a loner and will be happier as a single pet in later life.

Finally, during this sensitive period your kitten also needs to get used to human noises, household smells and ordinary human homes. He will learn that he need not be frightened by the noises of the washing machine, the telephone, the radio and the TV. The ideal home for a baby kitten is a noisy household with children, friends who visit, a calm and loving dog, and one or two other cats. Growing up in this atmosphere will mean your kitten is used to, and not frightened by, most domestic activities.

KITTEN TALE: Tiger and Lily were two feral tabby kittens found in a rubbish dump and rescued by Cats Protection (CP). The first eight weeks of their life had been without human contact, so they were more like little wild animals than pets. For a month they were kept in the home of a CP volunteer to train them into domesticity, then they went to the home of their new owners.

'The kittens hid behind the tall fridge in our utility room for about a fortnight and would have nothing to do with us,' recalls John. They ate and used the litter tray only when nobody else was in the room. Like wild animals they were frightened of humans.

'One evening my wife, Judith, practising the advice that the kittens were inquisitive, lay down on the tiles floor of the utility room and after a short time they came to inspect her, eventually crawling all over her.'

John had been warned that the brother and sister were particularly frightened of male voices. 'They got used to Judith quickly, although it took at least a month before they would trust me to feed them or be in the same room.' As they grew up, both cats began to be more relaxed around their owners.

But the lack of human contact in the crucial years of their kittenhood means that they only trust Judith and John, their son and a cat sitter who looks after them when the family are away. 'They are still deeply suspicious of strange voices and sounds, and a knock on the door or a ring on the bell sends them rushing for the exits,' says John.

Socialize to prevent a dysfunctional kittenhood

A dysfunctional kittenhood leads to a dysfunctional cat. Being a pet has to be learned. The worst upbringing for a pet cat is to be born in the wild. Kittens that are rescued off the street need a lot of handling as early as possible by their rescuers to overcome their fear of humans. Wild animals can sometimes be tamed by endless human patience but they rarely make easy pets. The other bad upbringing is that of a pedigree kitten reared in a cat chalet, a shed or a quiet spare room without enough human daily life around them.

Some pedigree breeders who love and cherish their cats simply don't understand the importance of socialising their kittens by exposing them at an early age to human activities, noises, smells and sounds. Other less scrupulous breeders will show the buyer their kittens in the house, but then stick them back into the shed when the would-be buyer has left. So though your kitten has a posh pedigree, the pedigree alone will do him little good. Only a good education and lots of handling will make for a happy pet.

If your cat gets pregnant, you must give the young kittens the best education. 'If the mother cat is happy about it, you can handle kittens carefully and gently from day one,' says Kathie Gregory, a cat behaviour counsellor and pedigree breeder of silver-spotted British Shorthairs. 'It's an old wives' tale that cats will eat their kittens if you do that – though some cats are more protective than others. Why wait until later?'

As a cat behaviour counsellor, Kathie Gregory has met many cats whose later life is badly affected by a poor upbringing as a kitten. Many of them were kept relatively isolated in the first three months of their life; cats that are nervous in later life may not have met enough different people while they were young. Cats that are aggressive to other cats may never have been in contact with other adult cats when they were kittens.

Kathie Gregory feels that kittens will benefit from human contact as early as possible rather than waiting for the age of three weeks old. A rescued stray mother would be best kept separate in case she is carrying a disease, but in normal circumstances Kathie Gregory recommends letting mother and kittens meet the other cats. In a household with a sensible number of friendly, vaccinated and healthy cats, this contact has very little risk of passing on disease.

'After two or three days, or as soon as the mother is happy about it, the other cats should be able to come and visit. My other cats are always in contact with the mum and kittens, and it socializes the babies to other cats.' Kittens that are kept separate from all other cats may find it difficult to be happy in a multi-cat household and would probably be happiest if homed as single cats. 'They are inexperienced with other cats and don't know how to cope,' Kathie believes.

Bringing up kittens in an ordinary human home, rather than in a cat chalet at the bottom of the garden, prepares them for later life. By three weeks old, they are happily crawling around and can be allowed into the house under human supervision. Remember to keep lavatory seats closed and access to any other dangers, such as chimneys, blocked off, though. Kittens can squeeze into all kind of unusual places.

By being around normal life, the babies will get used to the routines of cleaning, dishwasher noises, TV, and cooking activities, but introduce them gradually to really scary things. 'I don't vacuum with them in the same room until they are about four or five weeks' old and proficient at running,' says Kathie Gregory. 'I start by vacuuming in other rooms so they can hear it, and so they are more used to the noise. Then I vacuum the opposite end of the room to them, and build it up gradually, so I am vacuuming around them.'

She makes sure her kittens meet plenty of different people, not just herself and her partner. 'The kittens are usually booked by the owners before they are born, so I encourage the new owners to visit and handle the litter, as long as they wash their hands first. I invite the local kids round. In an ideal world I would like my kittens to meet people of both sexes, different ages, and different ethnic backgrounds.'

Food, litter and early learning

If you want your cat not to be too fussy about his food, you need to feed correctly from the start. At four weeks a kitten starts eating solid food and learns from his mother what he should eat. Researchers discovered that, if a mother cat had been trained to eat bananas, her kittens learned to enjoy bananas too. So a kitten that has been brought up on only one type of food will often prefer that type in later life. A kitten that has been offered a variety of different cat foods will be less fussy about what he will eat as an adult.

Kittens learn to hunt by watching mum. A mother cat starts bringing her kittens dead prey when they are about four weeks old. Later she brings in live prey and helps them hunt it down. Most kittens brought up in domestic homes don't eat wild prey. However, the hunting instinct is so strong that kittens that have never come across living prey can nevertheless learn to hunt even late in life.

Finally, there is the weaning experience. As the kittens eat more solid food, they begin to grow teeth and their mother is less keen on being suckled. When they are about seven weeks old she will begin to refuse to let them suckle. Thus, by being refused milk, the kittens learn to eat mainly solid food.

Preferences for a particular sort of litter in their litter tray start early. Kittens begin to use the litter tray around the age of five weeks. If they are placed on litter they will begin to dig it, and they get used to the smell and the feel of a particular litter under their feet. If your kitten was introduced to wood chips, he will probably prefer wood chips as litter when he is a cat; whereas a kitten given sand will prefer sand. The familiar kind of litter, indeed, will prompt him to go to the toilet.

So when you take your new kitten home, give him the litter that he is used to, even if that litter is not the one you prefer. You want your kitten to be house-clean from the start. For at least the first two weeks, stick to the old type of litter. During this period your kitten is getting used to a new litter tray in a new location – don't hurry him. Then, if you want a different kind of litter, change it gradually over about a month. First add a handful of the new type, then another one and keep doing this little by little so that after a couple of months almost all the litter is the new kind. This slow changeover takes patience, but it does help your kitten to stay clean in his toilet habits during the process.

KITTEN FACT: Weaning teaches kittens how to tolerate frustration. As they are pushed away from their mother, they experience times of frustration when they can't get what they want. During weaning, the mother leaves her kittens or lies down hiding her tummy. She starts refusing their overtures and they are left to cope without her help. This experience of frustration is important. It will enable them to tolerate other frustrating experiences later in life without losing their emotional cool.

Bottle-fed kittens sometimes miss out on this frustration experience. Even if the bottle is withdrawn, soft food is immediately offered in its place. They never experience a refusal and, rather like spoiled children, may grow into demanding cats who cannot tolerate frustration. These cats use their claws on humans to get their own way. So if you are adopting a bottle-fed kitten or bringing up a baby on the bottle, don't be afraid to gently and lovingly push him away if he is being too demanding.

What you can teach your kitten

Although the most important lessons take place before the age of eight weeks, there is a further seven weeks in which the growing kitten can still be learning. These are the weeks that will partly coincide with the beginning of his life in a new home, so it is therefore important to influence your new kitten's behaviour as early as possible.

If you start early enough, cats can be trained to accept all kinds of lifestyles. If you live in a caravan, or even a boat, and move from place to place, a kitten will accept this as part of his normal life, as long as he is introduced to this way of life from an early age. If you have two homes – a town flat and a weekend country cottage – you may want a cat that is not frightened by car travel. In this case the best plan would be to get a kitten as near the age of eight weeks as possible and accustom him to being driven by car from the beginning. If you explain your situation to a rescue shelter, they may help by letting you take home the kitten early.

Many pedigree breeders will not let a kitten go to his new home until he is 12 weeks old – the age recommendation that the various cat organizations have laid down. Again, if you want the cat to be able to travel in a car, you might ask the breeder if she will let you take him for car trips before you bring him home. If he refuses, you will just have to accustom him to car travel as soon as possible. To do this, take him for car trips every day for the first two or three weeks. The experience should be made as pleasant as possible. Small items of particularly delicious food should be offered when the kitten is first put in his carrier, and the box can also be sprayed with Feliway® (a synthetic copy of cats' facial pheromone that they use to mark their territory), which promotes relaxation in cats.

Indeed, all cats should be encouraged to love their carriers and a few little tricks make this easier. Leave the carrier in full view in the house. Place small portions of food in it a couple of times a week, to be discovered when you are not around. Make the inside cosy with fleece. Also consider introducing a crate into the house and make it into a kind of cat den; then, if you are ever visiting relatives with your cat, you can take the crate with you as a home from home.

KITTEN TIP

Never buy a kitten on impulse and always check out the breed to see if it has inherited diseases. The Feline Advisory Bureau (www.fabcats.org) has a list of these diseases and holds registers of breeders that screen for these disorders. Kathie Gregory, a cat behaviour counsellor and pedigree breeder of silver-spotted British Shorthairs, has compiled this list of what to look for in a breeder:

- The house should not be overcrowded and the breeder should have enough time for the cats he has. Beware of kitten farms advertising 'Kittens always available'. Good breeders breed one litter at a time.

- The breeder should be concerned where their kittens go to live and should ask you questions to see if you can offer a good home.
- The premises must be clean and tidy and the cats and kittens healthy and bright-eyed, with shiny coats. Adult cats must have been vaccinated.
- The cats and kittens should respond to the breeder and not shy away.
- You should be able to see all the kittens running around with mum. Beware of anyone just bringing one kitten to you to view.
- The breeder will encourage you to handle them, but ask to wash your hands first if you have been in contact with other cats.

'Don't worry if kittens squirm and won't be handled by you,' advises Kathie Gregory. 'I've yet to meet a kitten that was happy to be held and would keep still when there is so much to play with! It will be obvious if kittens are fearful, rather than just not interested in being held. Check out mum: if she's friendly, the kittens are more likely to be friendly too.'

KITTEN FACT: Cats, like humans, are influenced not just by their education but also by their genes. Scientists have discovered that a kitten's temperament is influenced by the temperament of his father. Nervous or aloof fathers sire nervous or aloof kittens. So some kittens will grow up to be more nervous or aloof around humans than others. You can educate a nervous kitten to become a cat that is confident around humans, but you cannot make him into a cat that enjoys cuddles.

Scratching – it's a post or the furniture!

Cats need to scratch; it's what they do to condition their claws but, just as important, it's what they do to mark their territory and make it smell right! Remember, smell is home to a cat. Cat-loving homes often show visible evidence in the form of torn furniture, wrecked carpets and frilled curtain ends. To avoid this damage, install a cat scratching post from the beginning and give your kitten plenty of attention when he scratches it. If he is slow to use it, place small items of choice food at the base. Don't ever force his paws into contact with the post but, instead, play with a bit of string round the post to encourage him to make claw contact.

Nail clipping and teeth cleaning should also be started as early as possible and performed frequently. Even if you don't need to, look into your kitten's ears and mouth every week and pretend to clip his nails. Long-haired cats will need grooming almost every day, while short-haired cats will benefit from a weekly groom when they are older, so get your kitten used to brushes and combs all over his body. The younger your kitten, the quicker he will become accustomed to these. Each experience should be made pleasant with good food treats.

Don't let your cat treat you like a mouse

Stop bad behaviour happening from the beginning. Do not let your kitten play rough games. It is so charming to see him pouncing with his claws out on your finger, and so tempting to wiggle your fingers to make him play with them, nibbling them while inflicting a tiny bite; the wound is so small that it hardly seems to matter at all, but when your cat is an adult and he continues the same rough games, the pain can be quite intense. So it is important that from the earliest days games using claws or teeth are not allowed and the fun stops immediately at the slightest use of claws or teeth. Kittens must learn that playing is only acceptable if the claws are sheathed.

Ambush games should also be discouraged. It is easy to laugh when we see a tiny bundle of fluff hurl itself against a large human being with its tiny claws out, but when a full-grown cat dashes from the bushes and rakes your legs with his claws, it is no fun at all. Ambush games are particularly delightful for cats since they are based on the predatory sequence – the basic instinct which ensures that performance is always rewarding. If you are not careful, you will find that you are being hunted by your adult cat as if you were a mouse, so it is vital not to let your kitten play ambush with you. Each time the little creature bounds towards you, turn your whole body away and ignore him completely.

Cats don't learn from punishment

Human beings mistakenly believe that the best way to get rid of bad behaviour must be punishment. Dogs may lick the hand that hits them, but cats won't. Punishments that produce pain or fear will totally poison your relationship with your cat. Remember, whenever possible, cats avoid danger before it happens, so rather than risk punishment from you, your cat may just decide you are not a safe person to be near. Worse still, your cat might just leave home in order to find a core territory safe from what he considers abusive human behaviour.

Your cat (like all animals) will respond much better to rewards for good behaviour, but to succeed, you have to learn what is rewarding to your individual cat. What you think is rewarding is not necessarily what *he* thinks is rewarding. Petting is very important to us, but usually less so to a cat. Dogs will work for human approval, but for a cat verbal praise will not be very motivating.

Cats don't much care whether they please us; what they do care about is getting our attention. They love making us focus on them. As well as food, attention and games are good rewards and individual cats may respond best to just one or two of these. There's no need to punish; there is an alternative – withdrawing attention and all food rewards. No pain is inflicted, only disappointment. If a kitten is expecting attention and does not get it, he will learn to avoid the behaviour that meant he was ignored, even shunned, by his human!

For instance, if a game with your kitten gets rough, stop the game, let the kitten fall from your lap (if he was there) and walk away. If you can't walk away, turn your whole body away, giving no eye contact and keeping absolutely silent. The kitten will learn to stop clawing because claws end the game and lose your attention. Continuing the game would be a reward. Ending the game and withdrawing attention disappoints the kitten. Games with correctly sheathed claws can be encouraged by the continuance of the game – a reward. The message is: sheathed claws, good; unsheathed claws, bad!

Learn to think like a cat

Changing cat behaviour is difficult. We humans find it difficult to think like cats and we also don't have their persistence. We don't time our rewards fast enough. We think a cat understands us far better than he does.

For instance, people often say 'Bad cat' in a threatening tone of voice to a cat that is scratching the furniture. From the human point of view this is a punishment, but a confident cat may well think of the rebuke as a reward – he has successfully got his owner's attention. He stops scratching temporarily because now he has achieved his aim – human attention – but he will scratch again when he wants his human to notice him again. Scratching achieves his aim.

Training a cat to do 'tricks' is far more difficult than training a dog. It has to be done with rewards because punishments wouldn't work, and it also requires human skill and patience. But it can be done and it can be very enjoyable, particularly for an indoor cat. Your indoor cat will love jumping over an activity course, offering a paw, sitting up on his hind legs, or lying down. Good trainers, usually using clicker techniques, have taught their cats (at home, not for public performance) to strum the piano, to open doors, or even to meow to music. There are now easy how-to books about clicker cat training. Your cat will let you know when he has had enough clicker work, by simply walking away.

Sex – get your kitten fixed

Sex rears its ugly head at the age of about five or six months. Cats under a year old are not fully grown but they can still get pregnant. Don't delay: if you want to give your cat a long

KITTEN TIP

When getting your kitten neutered, remember to have a microchip put in at the same time. A central agency holds the microchip details and your address. It means that if the kitten wanders off and is handed into an animal rescue shelter he can be identified and the agency will contact you so you can get him back.

and healthy life, neutering and spaying is one of the most important benefits you can give him or her. This operation is traditionally performed between six to nine months of age, though some animal shelters do it earlier.

The idea that every female cat must have one litter is just outdated folklore, and so is the one that neutered male cats are somehow missing out. (This theory is sometimes put forward by men, who are thinking from below their navels.) It is because humans believe in these outdated ideas that there are tens of thousands of homeless unwanted cats and kittens on our streets and in our countryside.

As I've already mentioned, a male cat who is 'entire' becomes a smelly wanderer rather than a home-loving pet. Entire males go off for miles in search of sex; they hang round street corners caterwauling and fighting each other. They can catch life-threatening diseases from this behaviour, or just get run over, and they may also compulsively advertise their masculinity by spraying on vertical surfaces inside your home. You don't want your cat to become one of these.

Female cats can have three litters a year, and rescue pens are full of pregnant strays – thin and small females who are exhausted from repeated kitten bearing. In some countries many of these stray cats will end up in shelters that will kill up to seven out of ten of the cats handed in. Even no-kill shelters, like Cats Protection in the UK, risk chronic overcrowding in the kitten season. Anybody who breeds from a cat without a good reason is perpetuating cat overpopulation.

So, get your kitten fixed.

CHAPTER FOUR

When an Adult Cat Adopts You

'Cats have the credit of being more worldly wise than dogs – of looking after their own interest, and being less blindly devoted to those of their friends. Cats certainly do love a family that has a carpet in the kitchen more than a family that has not: and if there are many children about, they prefer to spend their leisure time next door.'

Jerome K. Jerome (1859–1927)

It is flattering to be chosen as a lifetime companion by a cat. Kittens are enchanting, but if you lead a busy life you can't really give them the proper education that they need. Adopting an adult cat is a good way to choose the right sort of cat that fits your lifestyle. You miss out on some of the fun, but you also miss out on retrieving your kitten from the oak tree, losing your socks to a kitten who has poked them under the bed, tripping over the plastic bottle used as a toy on the kitchen floor, or explaining to your neighbour why you were seen crawling round the garden in the dark with a torch calling 'Kitty, kitty'.

A surprisingly high number of people do not choose a cat; she chooses them. Sometimes I think the cats of the neighbourhood leave scratch marks in code near garden gates which say, 'Here's a sucker. Move in fast.' Most cat lovers, during their lifetime, will find they acquire a cat not because they want one or need yet another one, but because one just moves in.

CAT TALE: Jimbo was the churchyard cat in a Northamptonshire village. He used to greet members of the congregation and escort them up the long drive to the church door, returning to the gate to escort the next ones. He also would sit companionably near people who were tending their family graves. Most people thought he lived nearby and just enjoyed the churchyard as part of his territory.

Luckily for him, one of the congregation made some inquiries and found that he didn't seem to have a home. 'I went to see if he was all right after a bad February storm and discovered him in a poor state under the yew tree,' said Jean. 'I thought he was a black bin liner at first and he was so deeply asleep, so exhausted, that he didn't stir.' She took him some food and with the help of a cat-loving friend took him home in a cat carrier.

Jimbo had been living rough and his friendly behaviour to the congregation was the action of a lonely cat. If Jean hadn't bothered to look out for him, he would probably have died from starvation and cold, for he was eight years old, no longer a young cat. Despite the local paper running a story about 'the churchyard cat', his original family never reclaimed him.

His behaviour in his new home was that of a frightened cat. 'If a hand went out to stroke him, he would snarl, even though he had never done that in the churchyard,' said Jean. 'He wouldn't be nursed. But after two years he suddenly jumped up on my grandson's lap and gradually he has become very loving. He waits for me when I go out to the post box and comes to meet me as soon as I turn up in the lane.'

At the age of fourteen, Jimbo has, like many cats, started bossing his owner. He reminds her when it is bedtime and settles down on the bed, just at the spot where the electric blanket is warmest. He is very vocal and has developed the elderly yowl found in many older cats. 'If I sit down in the armchair he gets up onto my lap and keeps bending his head back to look up into my face for reassurance.'

Jean has been his saviour and he is happiest when he is close to her.

When a cat chooses you

Sometimes the new cat is a stray that starts hanging round your garden. Of course, you don't want a cat. Naturally, if she looks hungry, you leave some food down occasionally, but you don't want a cat. Then you start leaving food down more often, though you still don't want a cat. She starts turning up regularly for her meals, though you don't want a cat. You can't really refuse to feed her – though of course you don't really want a cat. Perhaps if you let her into the house, you could take her to the local animal shelter after a day or two, because you don't want a cat.

Predictably, that is not how the story ends. You don't really want a cat but you now have a cat. The cat has won her new home.

Some of these pathetic stray cats look as if they are wild and feral when you first meet them; they won't come near even when they are desperate for the food you are putting down. They have learned the hard way that many human beings will throw stones, shout, take aim with an air rifle, or even poison cats. The worst risks to a stray cat are torture by gangs of youths, being thrown alive on bonfires, or just being kicked to death. No wonder they have lost their trust in human kindness.

CAT FACT: If cats are living rough without much food, they can postpone growing. Gracie was a small thin black and white flea-ridden cat that turned up at a community centre. After her second visit one of the community workers took pity on her and took her home. In the next six months, Gracie doubled her weight and increased her size by a third. From a small cat about the size of a normal three-month-old kitten, Gracie grew into a large and slightly tubby adult.

The moment of truth, when a cat decides that this human being is a loving being, may come suddenly. After days, or even weeks of fearfulness, the cat may suddenly jump up on your lap or rub against your legs.

Once they feel safe again, stray cats will repay your kindness with a touching gratitude. So, having the patience to adopt a neighbourhood stray cat, as long as it once had a home, will usually work out well if you are the kind of person who accepts a cat as she is, not as you wish her to be. Some stray cats may always be nervous or too frightened for a cuddle; even so, most cat lovers feel it is an immense privilege to be chosen, rather than do the choosing.

CAT TIP

Before taking in a stray cat, you may want to know if it already has a home. Cut a strip of paper the size of a collar and write on it: 'Please phone this number.' Attach the paper as a collar round the cat's neck, using sticky tape. If the cat already has an owner, they will call you. The collar will tear off easily if it catches on a branch.

CAT TALE: When Liz moved onto a Thames barge moored at a boatyard in London, she was adopted by the boatyard cat, Samuel. A tabby, he was living on the next-door boat but when his owners moved, they just left him behind. Samuel very sensibly adopted Liz.

'He refused to set foot on the old boat, where he had been abandoned, even though he's completely at home on boats,' she told me. 'He visits every other boat, but not that one. He's really the yard cat now and he chooses where to sleep.'

Liz works a night shift and goes to bed in the early morning. She will often find Samuel waiting for her in the car park at 6 a.m. Then he joins her under her quilt, sleeping alongside her, elongated to fit into the relatively narrow bed, in the afternoon.

'He's a real toad! If it's cold, he's there with me rather than out hunting!' she says. 'He wakes me up by bouncing things like crab apples on my head.' At 9 p.m., before leaving for work, she feeds him and he walks off to find amusement elsewhere – either hunting in the yard or visiting other boats.

Adopting a feral cat

Truly wild cats, the feral ones that have never been socialized to humans, may never become good pets. The kindest way to adopt these cats is to make them into outdoor or stable cats. Cut a cat flap into a shed or outbuilding where the cat can find shelter from bad weather and put down food regularly. Ask your local cat rescue organisation if you can borrow a trap, so that you can get the cat neutered or spayed– for his or her own good.

The relationship between man and feral cat is rather like the relationship between a bird lover and the robin in the garden: the human provides food and the bird enjoys it. She may even become quite tame, but remember that she is a tamed wild bird, not a pet bird.

Feral cats can sometimes turn into household pets; over years, not weeks, they can be tamed. However, usually they will only stay close to their immediate human family; if strangers or visitors arrive, they will make themselves invisible. They are tamed rather than tame.

Taking on a feral cat with the idea of making her into a household pet, is not for the faint-hearted or the impatient. A former feral will never be cuddly and may never sleep on your bed at night. Ferals are only for cat lovers who enjoy a challenge and have infinite patience.

CAT FACT: Britain's largest cat rescue charity, Cats Protection, rescued 55,000 cats and kittens in 2007 and helped neuter 121,000. They also gave 16,800 feral cats a new start, by neutering them, treating them for fleas, and then finding them a home in stables or farmyards.

Taking in a rescue cat

Adopting a rescue cat is a brilliant idea for anybody who truly loves cats. There are thousands of cats and kittens that desperately need homes, and usually the kittens are the first to go. The older cats are the ones that get ignored or left for months in their pens, waiting for someone to love them.

First, choose your rescue charity. There are still rescue organisations that are little better than cat hoarders; you may be taken into a house where there are up to 60 cats on every surface, or you may be escorted round a small holding where cats are held in dirty wire-netting runs, locked up in the equivalent of rabbit hutches, or where scores of them are simply allowed to roam around. By all means rescue a cat from these horrible surroundings, but after you have done so, report the owners to the local RSPCA. There is no excuse for charities to hold animals in bad conditions.

If you find yourself at such a rescue shelter and feel you must take home one of the poor cats there, be prepared for a big veterinary bill. Also consider proper quarantine in the spare bedroom until you are sure your new moggie has been restored to health. If you already have a cat, you don't want the new arrival to spread disease to her.

Good rescue shelters will quarantine new arrivals and keep their cats in separate pens. The surroundings will be clean and the staff pleasant, and they will also insist on checking your home surroundings before handing over a cat. Any rescue that just lets you take away a cat on your first visit is suspect. Home checking takes time, but it is a sign of good welfare. They will give you information about the cat's vaccination status and also whether she has any peculiarities.

Ask plenty of questions. Has the cat been neutered? Is the cat vaccinated? Does the cat have any health issues? Has she been tested for FIV? (Not all rescues can afford to test for FIV, but you will need to do so yourself if you are adding a new cat to a household where there is already one.) Does she get on with dogs? Does she get on with other cats? Does she have any behaviour problems? Good cat rescues will be able to answer most of these questions.

CAT TIP

If you have adopted a nervous rescue cat, buy some Feliway® spray from your vet and spray it in the room where she will spend most of her time, before she comes home. Or install a Feliway® plug-in diffuser and let it run for 24 hours before her arrival.

CAT TALE: ZigZag, now a handsome and healthy marmalade cat aged 12, was a kitten living in a house with 70 other cats. The house owner was an animal hoarder who never refused a cat. As a result, stray cats were regularly handed in to her and the numbers rose to an unacceptable level.

Somebody told Angie that she could get a kitten from a woman who rescued cats, known as the Mad Cat Woman of the town. And sure enough, when she rang, the woman told her that she always had kittens available.

When Angie arrived, she realized that this was a cat hoarder's household. There were cats everywhere – under the beds, behind the sofa, on every chair and table, and a conservatory full of cats sunning themselves. She was offered a basket of kittens and told to choose one.

Thinking that any kitten would be better off out of the house, she chose ZigZag. He looked healthy but she took him to a vet for a complete check. He had fleas, lice, and ear mites. A day or two later it was clear that he also had diarrhoea and the vet was worried about his breathing.

A long course of antibiotics and several more vet visits followed. ZigZag had been free except for a donation, but his initial vet bills in 1995 were £800. Worse still, Angie's other cat, Beatrice, had been put at risk of infection.

Households with scores of cats are not good places to find a healthy kitten.

How to choose a well-behaved cat

Most people just wander round a cat rescue place and peer into the cat pens. They see a cat they like the look of and feel drawn towards her. 'I fell in love with her,' is what they often say. Choosing a cat on looks alone is probably the worst way to choose a cat.

The secret of choosing the right kind of cat is to focus, not on looks, but on behaviour. 'Handsome is as handsome does' applies just as much to cats as it does to men. Ask to go into the pen and handle the cat. If this request is refused, find another cat rescue!

Cats behaving badly are common in rescue centres. Some people just get rid of difficult cats by dumping them on a nearby farmland. Some even throw them out of a car on the motorway. Others, the relatively responsible owners, take them to rescue centres with a made-up story. A few good owners take them to rescue centres and tell the truth – this cat bites, or this cat sprays in the house.

If you are choosing your first cat, or if you have young children or elderly or ill relatives in the household, you need to select a well-behaved cat. A cat that bites or scratches human beings isn't safe for young children or for people with poor immune systems. If this is your situation, ask for a friendly and cuddly cat. Go into the cat pens and see if the cat will let herself be picked up.

Some cats in rescue centres are so starved of human affection that they will throw themselves into your arms, climb on to your lap, and purr ceaselessly. This is the ideal cat for the household where somebody is in all day and where children will want to have close contact. Very affectionate cats are not everybody's choice. They may be tiresome for those who work at home – trying to finish a list of figures on the word processor is tricky when a cat is lying purring on the keyboard or snuggled under your neck grooming your ears.

The super-affectionate cat is also not a good cat if you are going to be out all day working. Left without human affection, she will probably find a neighbour who can give her the company she needs. You will end up with a time-share cat – not always a bad situation, but one you may not want.

Independent cats that just want to get out and get hunting are ideal for households where there is nobody home during the day. These cats have plenty to do and do not need or want too much human company. They tolerate rather than enjoy petting. They will come home for meals, particularly if meals are given at the time their owners return. They will also enjoy the warmth of the bedroom at night. It's never a good idea to let cats out at night, particularly not these committed hunters – they may fall prey to foxes or to cars, as in the dark they concentrate on their prey rather than their own safety.

So, when choosing a cat, do not make the decision based on looks or colour; select the temperament and behaviour that will suit your household.

CAT FACT: Black cats and black-and-white cats are less popular than tabbies or gingers, so they find it more difficult to get a new home. Indeed, in some countries black cats are just euthanized when they are handed into cat rescue shelters. If you want to do a kind act, choose black or black-and-white.

Choosing a badly behaved cat – yes some wonderful people do!

Most cats are anxious and unhappy in rescue pens, however well the rescue charity is run. Some nervous ones just cower in their cat bed, terrified of being stared at through the wire. You will often find it difficult to see these cats, as they hide as much as they can from people walking through the centre.

Yet they make rewarding companions for people who are patient with them. It may take them six months to a year to get back their confidence in the human race, but they will eventually become much-loved and much-loving pets. Seeing them change from timid frightened beings into confident family pets is a wonderful experience.

Other frightened cats are on constant alert, patrolling the edges of their small territory in the pen in an aggressive way. Many are very upset by the presence of other cats, and, alas, many cat-rescue establishments have see-through barriers between occupants. I have seen these cats try to strike out at the intruders they can not only smell but also see next door. These are not the ideal cats for a suburban house where there are many cats in the neighbourhood; they need a home where there are not scores of other cats nearby. However, when safely installed in a home where there are no feline competitors or intruders, they too will relax.

Some cats nip, but this is often because they are stressed and worried; these are the cats that are fundamentally still slightly afraid of humans because they may not have met enough of them when they were kittens. Gentleness and patience on the part of their new owner will reduce this habit, but they may never become entirely relaxed in close physical contact. They will sleep on the bottom of the bed, but never inside it or in your arms! These cats get nervous with too much affection and may lash out – more on that later (see pp. 96–109). A good rescue establishment will tell you about this in advance.

Finally, there are the confident biters, who use teeth and claws to discipline humans. They have learned that it works. They nip for attention, or they nip when a human stops petting them. They may even nip if a human tries to move them off the armchair. Confident biters can be rehabilitated more easily than nervous biters – more on that later, too (pp. 102–4, 157). Why choose a biter? I know people who absolutely adore their cat even though she bites them regularly. They find it funny that a small animal should do this to a large human. And so it is. Biting cats need all the friends they can get, so if you don't mind being nipped there are plenty of biting cats in rescue shelters who want a home.

Adopting one, two, or three cats?

Cats are not group animals like dogs, although some have a social life and make close friends. If you see two cats sharing a bed in the rescue pen, you can be sure these are either friends or relatives and therefore it is safe to adopt both of them together. Kittens from the same litter also frequently grow up to be good friends.

However, you will often see two cats in the same pen, but in different areas of the pan. One cat is outside; one is inside. One is up on the high shelf; the other is below as far away as possible. Rescue places often house cats together if they have come from the same home, but quite often cats from the same home are not necessarily friends. Even in the confined space of a pen, they keep as far away from each other as possible.

These are cats that can be adopted together but are not a true pair. You will probably find that, once they are in more spacious surroundings, they live in separate areas of the house. That is fine if you have enough space, but if you intend to keep them as indoor-only cats in a relatively confined area, these apparent, but not really friendly, couples are not the best choice. You should search for a pair of cats that share a bed together and are real friends.

Finally, never adopt one cat from one pen and one cat from another pen; you can have no idea if they will get along together. Some of them will; many of them won't. Besides, any rescue centre that suggests this arrangement is obviously ignorant of the risks involved. If you take home two unrelated cats, hitherto unknown to each other, the first six months are going to be tense. They may settle down amicably in the same household, or they may not. You cannot tell and you may find yourself taking one back to the rescue centre.

Above all, don't adopt too many cats. As a rough guide, there should not be more than two cats in a three-bedroom house. The greater the number of cats, the greater the potential for cats becoming unhappy and behaving badly.

Adopting an indoor-only cat

In the US, indoor-only cats, which are never allowed off the premises, are common. In the UK we usually keep cats in a house with a cat flap opening into the garden. Most UK rescues, therefore, insist that their cats go to homes like these, where the animals can have the experience of enjoying outdoor life. But even the more rigorous cat rescue charities will have some cats that are suitable for the indoor life.

Indeed, there are some cats that should not be allowed out. Blind, deaf or disabled cats cannot fend for themselves in the outdoor world, and so for their own safety they must be confined to indoor life. A surprising number of disabled cats turn up in rescue centres, perhaps because they have been unlucky in their previous owners or been injured by cars during the life of a stray. Rescue charities love people who will take on these cats.

The other category of cats that are indoors-only are the FIV positive cats. Feline Immunodeficiency Virus is a disorder that is not unlike HIV in humans, but FIV cannot be transmitted from cats to humans, so they make perfectly safe pets. However, the virus can be passed on to other cats, so it is to stop this happening that these cats must live indoors. Many of them have been picked up as strays on the street, where they caught the virus in the first place, and some cat charities will offer to pay future vet fees.

Elderly cats can usually live happily without the great outdoors; you will find them among the younger, more vigorous adult cats in a rescue shelter. Often they have a particularly lost look, as if they cannot understand how their owners could have abandoned them to the charity of strangers. Almost all cats suffer at least a little in rescue shelters, but the oldies suffer most.

Week after week the elderly cats are passed over for younger cats. As the months go by, their thinning hair and old looks just don't interest the general public. If they are slightly arthritic, as many are, they may find life in a cat pen not just lonely and frightening but also uncomfortable. The high shelves and the relative lack of heat are hard for them to bear, particularly those whose happy life came to an end when their owner died.

These lost souls will settle happily as indoor cats. They don't ask much. A warm bed to sleep on, enough to eat, and human company for their last years is all they require. And some cat rescues will pay their vet's fees too.

Adopting a feral cat for stables or farmyard, not for the home

Many cat-rescue organisations have feral cats that have been handed in to them. Finding homes for these wild animals is not easy, as the adults ones will not make good pets. If you have stables, outbuildings or a farm, adopting one or two feral cats makes good sense, as you will have excellent pest controllers to keep down the mice and rats and reduce rabbit numbers in the summer. They will also keep off other cats, so that you will not end up with a huge colony of disease-ridden felines. Most cat-rescue charities will greet you with open arms if you can help them find warm barns and stables where these cats can live.

CHAPTER FIVE

Loving Cat, Aloof Cat

'Calvin ... liked companionship, but he wouldn't be petted or fussed over, or sit in anyone's lap a moment; he always extricated himself from such familiarity with no show of temper. If there was any petting to be done, however, he chose to do it. Often he would sit looking at me, and then, moved by a delicate affection, come and pull at my coat and sleeve until he could touch my face with his nose, and then go away contented.'

Charles Dudley Warner, 1829–1900

Even in the love affair between humans and cats, the path of true love never did run smooth. We fall in love with a cat, lavish him with kindness, let him sleep on the bed, take the best chair, and sometimes he treats us merely with disdain. Cats are the natural Mr Darcy to our Elizabeth; their sense of superiority is only too clear, but somehow it attracts us.

Cats may have become part of the human family, but they are under no sense of obligation to us. Just occasionally there is a cat who loves too much, but more usually it is the human who loves too much! There's no question of cats looking up to us or obeying us; in their eyes we appear a clumsy and not very bright species, tolerated for our cat food and loved as a pet. Cats own humans more than those humans own them.

Many of us cat lovers collect scratch marks on our hands (apparently about six out of ten cats will occasionally scratch or bite people). Dogs may growl or bark at strangers, but cats are equally happy to hiss or scratch their own family! We put up with behaviour from cats that we would never tolerate in dogs, and yet we still love them. We have this small tiger in the home, that thinks nothing of clawing us when we offend, and yet we still love him. All those years of domestication have failed to crush the independence of our felines.

Of course, the scratchiness of cats starts at kittenhood. Kittens that have never been properly handled will remain half wild at heart. However, temperament is also hereditary, and it is possible to have siblings where one is friendly and one is aloof. Remember that ordinary moggies that go out on the tiles to find a mate often have a litter by more than one father. If one daddy is a nervous tomcat and the other daddy is of a laid-back temperament, their kittens will differ in temperament too.

As for those of us who take in stray cats or knowingly adopt difficult rescue cats, well we have to settle for what we have got. Most rescue cats, traumatized by life on the streets or even by life in a rescue centre, will become at least a little more friendly when they settle down to home life. Some may even blossom into extremely loving cats.

Can you force your cat to change? Wait a moment: does the idea seem familiar to you? Many women and quite a few men have decided that they can *make* a partner settle down. They get involved and try to change their partner, making them over into a different person. This doesn't work. That women-who-love-too-much thing may make you hope that you can change your cat, but be warned: you will have to change your own behaviour first, before you can change your cat's.

It is far better to start your relationship with acceptance. Accept how your cat is. Now work out how you are going to live happily ever after together.

Why can't my cat love me more?

The cat–human relationship is a bit of a mystery. It's as if we humans can never quite get our cats to love us enough. Or so I was thinking the other day as I tried for the hundredth time to persuade my tabby-and-white cat, William, onto my lap. I can lure him there with food and he will remain there, awkwardly sitting rather than relaxing on my knees, until the food is eaten. Then he jumps off with an evident expression of relief.

Why won't my cat sit on my lap or allow itself to be cuddled? I used to get asked this question over and over again when I was a pet agony aunt. There was a yearning desire in all the cat lovers just for more love. They wanted purrs, and rubs, and licks, and what they were getting was the sight of a cat's backside as, having finished his dinner, he was trotting out of the cat flap to pursue his daily round or taking himself off to his favourite place on the windowsill to nap with a clear 'Do Not Disturb' sign.

The correct answer to these questions was that the said cat had probably either not been handled enough as a kitten or was just of a nervous or aloof temperament, but I think what the unsatisfied cat lovers wanted was a magic potion. They were in love, but their loved one just took them for granted. It was like being married to a partner who comes home, eats dinner and then goes straight off to the pub or the local railway enthusiasts' group, or a Women's Institute meeting.

Some cats are just not very friendly by nature. There are cats that like company but are less keen on touch, and these are the ones that will sit a careful three feet away on the same sofa – just too far for you to stretch out and touch them. They will accompany their human down the garden and sit nearby while he works at his desk. They are there – but never close enough to pet.

This is a visible expression of what ecologists call 'flight distance' – the distance at which a wild animal will move away from a human. Very wild feral cats will have a flight distance of about 300 yards, running as soon as they sight a human.

A cuddly pet cat has no flight distance at all, but a slightly anxious pet cat will have the flight distance of, say, three yards. He wants to be near you, but will often start to edge away as you move towards him with your outstretched hand. He may even sleep on the bed but he will be near your feet – a safe distance away from your yearning arms!

The only comfort I can offer to wistful souls like myself, who dream of the day when their cat will leap into their arms for a cuddle, is that the relationship sometimes gets closer in old age. Little Fat Mog, a very stand-offish small black cat, was fond of me but never wanted to be too close to me. In her old age, however, she became wholly deaf and much more dependent upon me. She never became a lap cat, but she would sit near to me, purring loudly and gazing upon me with loving, slightly anxious eyes. In those eyes I realized, at last, that I was as important to her as she was to me.

CAT FACT: Two researchers, in the *Journal of Business Research*, claimed that intimacy was enhanced by the cat's soft fur, the way cats can jump onto laps, sleep in the bed and allow themselves to be carried around like young children. They concluded that the bond between human and cat was 'one of the purest forms of relationships between God's creatures'.

CAT TALE: James, a handsome black-and-white, neutered male adopted from a rescue shelter, was an aloof cat. He didn't seem much interested in giving or receiving affection. While he accepted the cat food and the warm place on the best chair, he disliked being picked up and would wriggle out of Marianne's arms as soon as he could. 'Look but don't touch,' was his motto. James was too busy to bother with humans most of the day.

His aloof attitude was upsetting for Marianne. She wanted a warm furry being whom she could pet. James was having none of this. He never purred or miaowed when she touched him. He didn't even sleep on the bed at night, being content to be downstairs on the best armchair. Because she really loved him, Marianne concluded she had to accept him as he was and stopped expecting affection from him.

This changed, however, when James nearly lost his life after being attacked by a lurcher dog. He crawled home half dead and was rushed to the vet. He endured six operations and wore bandages for weeks. When he was sent home, Marianne had to help him to the litter tray because he could hardly walk.

James's relationship with her was transformed. He would no longer shun her. Instead he spent as much time as he could on her lap, purring and even drooling. He began to miaow when he wanted affection. 'A prince among cats' is how Marianne described him.

Usually, we humans have to accept cats (and other humans) as they are. Trying to change them doesn't work. Extended veterinary treatment, however, can sometimes lead to change. 'Very shy cats, even wild-born feral ones, can sometimes demonstrate major changes in their tolerance of humans, if they are given treatment for serious injury or illness,' says Professor Peter Neville, founder partner of the Centre of Applied Pet Ethology. 'The social bonds forced on the animal because of the demands of physical care are established and then appreciated simply for the value that they have in terms of comfort and protection.'

The way cats love

Cats don't express their affection in the same way we do. If you watch two cats that are friends, they will come up to each other and sniff noses, briefly. They might then intertwine their tails and walk together for a little bit. Apart from time in the same cat bed, their affection is expressed in short, low-intensity bursts.

We like to pick up and cuddle cats, but for them this behaviour is far too intense and too long lasting. They hate being forcibly held so that they cannot break free when they wish. Some love-hungry humans literally pursue their cats round the house. This unnerves the cat even more and she moves into flight mode.

Cats fear humans that persist in harassing them. What they like in a human is the ability to ignore the cat, until the cat *himself* signals that he wants contact. This is why cats love people who dislike cats. People that hate cats have excellent feline manners: they don't go up to a cat and start stroking him, they just ignore him, until the cat itself jumps on their lap.

If you have a cat that doesn't seem to want much human contact, stop pestering him for the love he will not give. Do not pick him up, place him in your lap, or keep petting him. Do not gaze lovingly into his eyes (for cats a direct gaze is a threat). Play hard to get, even if you don't feel like it. Leave the cat alone and let him choose if he wants to be close to you.

Cats that can choose if or when they receive affection stop feeling so anxious. Sometimes, if the cat is truly unsocialized to humans, it may take a long time, but usually a cat that is left to choose will become more friendly. You may never have a lap cat, but you will have a cat that stays close to you and clearly loves your company. Your displays of affection must have a light touch, be short in duration, and always stop *before* the cat has had enough.

CAT TIP

Clicker training, which I recommend for biting bullies, also works with aloof cats. You can use it to train your cat to get up on your lap and to sit there waiting for his treat though it will need lots of training to get him to stay there long!

Cats that bite – the petting and biting syndrome

CAT FACT: A study of cat bites or scratches reported to a public health centre in Spain found that more than a third of the biting incidents were the result of the cat being frightened, 9% occurred during play, 16% as the result of the cat experiencing pain. Petting and bite syndrome accounted for 15% and transferred aggression for 16%.

Some difficult cats have wonderfully accepting owners. One of them is Sue, the owner of Roofer, a strong grey-and-white cat whose aggressive nature terrifies even his vet. When I visited Sue to meet Roofer, she warned me to stand back and only take a photo of him at a safe distance. 'Otherwise he will just go for you, Celia. He doesn't like having his photo taken.' She has learned that Roofer will bite her nose, drawing blood, if she holds him anywhere near her face. So when she picks him up (which she only does when absolutely necessary), she has to hold him against her body. By no stretch of the imagination could Roofer be described as a satisfactory pet. Living with Roofer is rather like living with a wild animal. Yet Sue loves him and takes care of him. 'Even though he is horrible and mean, we love him a lot,' she says.

Roofer is an exceptionally difficult cat but even more ordinary cats have their nasty moments. A typical scene occurs when they roll winsomely on their backs, squirming and looking at us as if to say: 'Tickle my tummy.' How sweet dear little Kitty is, you think. So you bend down to tickle the tummy and four paws with rapier-sharp claws grab your hand. Your hand is trapped and if you try to withdraw it a trail of 20 scratches will mark your skin with blood. Try punishing by tapping dear little Kitty on the nose? The claws dig deeper and she may even bite you too.

Welcome to the petting and biting syndrome. Cat experts tell us that this clawing or biting is the feline way of saying, 'Stop now'. As a message it is an extremely effective one. Very few humans continue trying to stroke, not least because they need to get their hands away before more damage is done.

So why do cats do this Jekyll and Hyde act?

They flip from affection to aggression because they are frightened. Many cats, except for the super cuddly, remain slightly anxious around human beings. In their eyes these huge lumbering beasts are clumsy uncoordinated creatures who might hurt them by mistake. Too much petting frightens them. The quickest way to get us humans to stop petting is to lash out in defensive aggression.

Actually, if we humans learned to read cats better, we might see what was coming. The tip of the tail starts lashing; the cat may stiffen; the ears go back and the pupils start to dilate. If we were cats, exquisitely sensitive to the nuances of body language, we'd see what was coming and stop petting early.

So can we train our cat into better behaviour? Petting is much more pleasant for humans than for cats. I'm not saying it is impossible to train a cat to accept more petting, but it will take months and months and it still might not work. It is usually easier to accept that our loved feline just doesn't like much petting. Accept the cat you cannot change!

CAT TALE: Biting Bertie came from a Cats Protection rescue centre, where staff warned adopters that he bit and clawed. Naturally, he had been slow to find a new home. His new owners, Richard and Barbara, liked a challenge and took him home nonetheless. He lived up to his name. He would seek their attention by rubbing against their legs, jumping on to the sofa besides them, and responding with purrs to being stroked on the head and under his chin. Then he would strike. His other habit was to roll on his back invitingly, then attack with all four paws. He drew blood regularly.

He also bit whenever he was picked up. However, if he was focused on eating, he could be stroked, so Richard and Barbara tried using food to train him to tolerate petting. If he allowed petting, he was rewarded with cat treats. If he started to look aggressive, the treats were withdrawn. Training a cat out of petting and biting is really, really hard work, but after three months Bertie was a bit less reactive. Barbara was able to pick him up without getting bitten if she approached him from the rear.

If Bertie had been a dog, he would probably have been taken straight back to the shelter. But Richard and Barbara were true cat lovers. 'All the pleasure he gives us far outweighs the odd bite or hiss,' said Barbara. 'We rather enjoy his incorrigible nature.' Biting Bertie had found his forever home.

Cats that bite – frightened cats

Most cats bite humans when they are frightened. A terrified cat has the choice of flight, freeze, or fight. Most of them choose flight. Fearful cats will run under the bed and refuse to come out for the rest of the day, or run up to the top of the highest cupboard and cower there. Cats are small animals and it's safest for small animals just to make a run for it if they feel threatened.

Occasionally they will react by freezing. You will see terrified cats in a vet's surgery; all they do is sort of huddle into their cat carriers, making themselves as small and immobile as

CAT TIP

You can train some petting and biting cats. Hold some small but tasty cat treats in the palm of your hand. You need to be able to close and open the hand with the treats inside. The cat is shown the treats and after a few seconds of petting give him one – if he has stayed calm. If he begins to scratch or bite, close your hand over the treats so that they are no longer available. Each time he is on your lap, close the hand as soon as he shows any of the signs that he might start biting or clawing. You are rewarding the cat for tolerating petting, and refusing the reward when he begins to show any kind of aggression. During this time the cat should be free to come and go at any stage, so that if he gets too anxious, he can end the training. Training sessions should be very short and take place only a couple of times a day. This technique requires weeks of total consistency and patience, so it can be easier just to accept that your cat cannot bear too much petting, learn the signs of an impending attack, and stop petting in time!

possible. You can almost see the bubble coming out of their head which says: 'I'm not here. I'm too small to be noticed. I am hiding.'

The final response to threat is to fight. We think of these cats as aggressive or even vicious, but actually they are frightened. Roofer, Sue's cat, came across to most humans as a fierce cat, always stalking round his territory and trying to fight off other cats. Actually, fighting and hyper-vigilance is the sign of a frightened, not a confident, cat. Sue, whom I consider the ideal cat owner, understood that Roofer was at heart a frightened cat, which is why she was able to forgive him each time he bit her nose.

Cats fight when they can't run away. Backed into a corner without escape, any cat will fight. I have seen a cat standing on a fallen tree trunk and holding off a fox. Her fur was puffed up and her tail was like a bottle brush, sort of humped up near the

CAT TIP

If you have a cat that is frightened of being petted, stroke him with long gentle strokes down the top of his back. See if he also enjoys being rubbed on the forehead. Avoid stroking anywhere near the tummy!

back then curving downwards to protect her backside. One claw was raised to strike. Her attitude made her look really aggressive and the fox thought better of it and backed off.

Terrified cats are a special challenge to owners. The secret of avoiding cat bites is, whenever possible, to always allow them an escape route. If your cat has rushed under the bed to get away from you, he will only bite if you pursue him and try to drag him out. Leave him alone and he will come out in his own good time.

We humans sometimes have difficulty in letting be. In this, as in many other ways, we can learn a lesson from cats.

Cats that bite – to bully their humans

Quite unlike the frightened cat is the cat that bullies his human with claw and order tactics. This kind of cat bites when he wants human attention – when the human is slow in putting food down, when the human tries to move him off the bed, when the human passes him in the corridor without acknowledgement, and sometimes even when the human wearies of petting and stops.

These cats have often been hand-reared as kittens. When a mother cat rears a kitten, she weans it by refusing to let it suckle her, and so the little kitten learns to tolerate being frustrated by his mum and grows up with self-control. Hand-reared kittens, though, never learn to tolerate frustration and many grow up to dominate their humans without fear of reprisal.

They also discover that biting works well. A sharp nip forces a human to stop doing whatever the cat dislikes. Biting gets attention (the only better attention-grabber is spraying!). Most of us humans don't understand that seeking attention means any kind of attention. When we shout, squeal with pain, wag a finger, or just look at a cat, all these actions are perceived as giving attention. If a cat wants to be noticed, he does something we humans think is wrong. Biting is a very good way to get attention. It works.

Cats that bite humans in order to get their own way are difficult to ignore, but ignoring is the only way to proceed. There must be no cries of anger or pain, and the owner must give the impression that she has not noticed the attack in any way. Owners of these cats should turn their gaze away, say nothing, and turn their back on the cat. Protective clothing such as wellies, cowboy chaps from a joke shop, walkers' gaiters, or thick trousers to protect the legs and ankles often help. If it is necessary to move the cat off a chair, either thick gloves or a broom covered with soft material can be used to push it softly off, despite its biting.

Some of these cats bite when their owner *stops* petting them – the exact opposite to the petting and biting syndrome described earlier. Petting must be put under the owners' control. A little bell can be rung or a mat placed on the lap to show that it is now time for petting. If the cat has jumped up onto the lap at the wrong time, before the bell is run or the mat is in place, the owner should simply stand up and let the cat fall to the ground without touching it.

Clicker training often turns round these difficult and over-confident cats. When they learn to respond to a human trainer, who gives good treats, instead of calling the shots by tooth and claw, they become very nice cats to live with. Clicker training reverses the relationship so that the human is in charge and the cat is not. Oddly enough, I think, the cat enjoys clicker training so much that he feels he has simply discovered a new training method to make a human give good food treats. Clicker training is a rare example of humans and cats being both in the winning role.

CAT TALE: Tonk was a large, blue-eyed, pedigree cat in a lovely glowing pale orange with ginger head, paws and a ginger-striped tail. He may have been a pedigree Tonkinese. He was found abandoned in a hedge near a vet's surgery and taken in by Cats Protection. When he was first picked up he seemed an excessively affectionate cat. Tonk really adored being petted. When I went into his rescue pen, he would walk straight up and climb on to me, getting as near to my face as he could. He was, so to speak, an in-your-face cat. I have never met a more confident feline.

He could take any amount of petting but trouble started when I stopped the petting. He would be on my knee purring but if I simply stopped stroking him, he would turn round and give my hand a sharp nip. If I picked him off my knee and put him down, he would nip and, as I turned to leave the pen, he would nip my ankles again. He was a big cat and it hurt.

This habit of getting attention by biting was probably why he had been abandoned in the first place. He bit me. He bit the rescuer who looked after him. He bit the Tonkinese Rescue woman who came to see if he was a Tonkinese several times. Not surprisingly, she went away saying he was not a Tonkinese! Tonk could have been trained to stop biting, but not in his rescue pen. Instead, he was moved to a new and bigger rescue centre where he found a home six months later. I gave a suggested programme of training to the new rescue centre.

The principle was to stop all petting and walk out of the room as soon as he bit. Tonk would be given the petting he craved, but only when his owner wanted to do so. He would be shown a signal to tell him when petting was available. This could be a special piece of cloth, put on the knees, so that he knew when it was time for affection.

With consistency, he would learn that biting never got attention, and that he was given attention when his owner chose and put out the petting cloth to show so. I think clicker-training Tonk would also have helped him become a nicer cat (see p. 157). I don't know if his new owner ever got my suggestions. I hope she did, because if she didn't, I bet she's getting bitten!

Cats that bite – you are a mouse to them!

Bored cats with no opportunity for proper hunting will sometimes treat their humans like prey. This isn't ordinary aggression; it is the predatory instinct at work. Predatory aggression is regulated from a brain area quite different to the part of the brain which controls ordinary aggression. If the eye, stalk and pounce hunting sequence isn't exercised on mice, any moving target will do – such as a piece of string or a human running upstairs. As these cats spring on their human, with claws at the ready, the human lets out a shriek. This is the equivalent to a mouse's squeak to a cat – highly exciting, making the game even greater fun.

Hunting the human is often found among indoor cats that don't have a chance to mouse. The method of dealing with it is similar to the way of dealing with a bullying cat. All hunting attacks must be ignored – give no reaction and no squeals of pain (protective clothing may be necessary protection to stop involuntary cries of human anguish!). The moving human must stand stock still so that the fun of chasing a moving object ceases with immobility. Thus the cat learns that chasing a human merely results in human immobility. Stopping and starting like this is quite difficult to do, but then changing cat behaviour *is* difficult. A cat behaviour counsellor can help here.

In addition, a cat that starts hunting its human badly needs alternative activities. The cat's predatory instincts should be fulfilled some other way – either by encouraging it to get going on the local mouse population outside or by lots of games with pieces of string. (There is more on this in Chapter 7, pp. 129–57.)

Cats that bite – in the wrong place at the wrong time!

If you have ever intervened in a cat fight, you probably got bitten. I have been unwise enough to do this and received a very deep bite! Fighting cats are so fired up that they will bite any intervening humans without thinking twice. This is what is called transferred or redirected aggression. The cat isn't angry with you, just with the other cat that it is fighting. But he will turn and bite you if you get in the way.

A similar effect can happen if, for instance, a cat is looking out of the window and lashing his tail at some dog that is passing by. If you pick up the cat at this moment, he is likely to claw or bite simply because he is already in fight mode. Redirected aggression is sometimes the reason why two household cats that have got on well with each other for years suddenly fall out. The story of Tanya and Amber, who quarrelled badly, started in this way (see pp. 192–3).

Cats that slobber and knead

At the other extreme from cats that bite are the cats that slobber all over their humans. These are cats that have a kitten–human relationship. They will leap onto the human lap and settle down to knead their human's knees with their sharp claws. Several cats will do a little kneading, but these cats will continue for hours – if allowed.

At the same time, they may also slobber with ecstasy. For thosc of us with relatively aloof cats, this sounds gratifying. These cats seem so very loving. The accompanying purring is a delight but the pounding of sharp claws and a fountain of saliva are less enjoyable. Occasionally these kitten–cats will start sucking – ears, sleeves, or fingers.

This, of course, is the behaviour of kittens when suckling from their mother. Kittens purr and knead with their paws and salivate as they suck her milk. Instead of growing up, these cats have made their humans into a cat mummy, whether the humans like it or not. Most cat lovers like this behaviour some of the time, while finding it irritating at other times, so prevention is difficult because we tend to react inconsistently.

Is this behaviour that should be changed? I think not. Instead, put petting under your own control, using a similar method as the one used for cats that bully their humans. Find a piece of blanket to use as a knee mat, specially kept for petting times. The blanket on your lap is a signal that this is a time for love and affection; if your cat jumps on your lap when the blanket is *not* on your knees, stand up, without handling the cat, so that he has to jump down again. Only let him jump on while the blanket is in place to protect you from the sharp kneading claws.

If you can be consistent about this for several weeks, never letting the cat stay on your lap without the blanket, he will learn that the blanket is a signal for petting sessions. He will also learn that when you do not have the blanket on your lap, there will be no petting.

Is it worth bothering to train away licking and sucking? Retraining requires weeks of patience and complete consistency, which most people find hard to achieve; just one or two inconsistent moments are enough to ruin the retraining. A little sucking and licking does no harm, so I think it can just be accepted, though not encouraged.

A cat that starts exaggerated licking and sucking, however, should be discouraged. As soon as the cat starts sucking, stand up and let the cat fall from your lap. As long as the cat does not lick or suck, he can remain there being petted. Again, the cat learns that sucking and licking means no more petting.

The cat that loves too much is usually not a major problem; most experienced cat owners count themselves lucky to have such a loving feline.

CHAPTER SIX

The Cat Flap Cat

'If the cat waits for long hours, silent beside the crack of the wainscot, it is for pure pleasure. Cats do not keep mice away; it's my belief they preserve them for the chase.'

Oswald Barron, 1868–1939

The way we keep cats varies with human culture, rather than with cat requirements or cats' choices. If cats could choose, I think they would opt to enjoy both the warmth of the family home and the excitement of hunting mice outdoors. They do not think like humans, so they would not be able to assess the true dangers of outdoor life – traffic accidents, death from dogs, foxes, badgers or, in the USA, even coyotes.

Most cats in the UK have an indoor/outdoor lifestyle, coming and going much as they please. Indeed, many UK animal shelters require new owners to let their cats have an outdoor, as well as an indoor, life. While there are exceptions, such as elderly or disabled cats, there is a perception in Britain that locking up a cat indoors all her life is cruel. Cats that can

CAT TIP

When you move house, choose the right home for your cat, if she has the freedom to go out of doors. Homes near busy roads are dangerous, particularly if there is an enticing bit of wasteland or field or wood on the other side. Even smaller roads can be dangerous if they are used as rat runs for cars during the rush periods.

freely go out of doors may live less long than those shut indoors, but they live more richly and in the UK many of us feel they live more naturally. They are also less likely to develop behaviour problems than indoor-only cats.

In the USA, and now increasingly in Australia, however, there are many more cats who live only indoors. Indoor-cat enthusiasts believe that allowing a cat to run the risk of death by car or predator is cruel. They point to the cats that get lost and end up homeless, those pedigree beauties that are picked up and stolen, the cats that catch not just fleas but fatal infectious diseases from another cat, those that are attacked and killed by predators, or those unlucky feline victims, probably only a small proportion, who are tortured or killed by humans who enjoy cruelty.

A century ago, the idea of keeping cats indoors all the time would have seemed strange indeed. In the stately homes of Britain three generations or so ago, animals had their different places. Pet dogs were kept upstairs with the family, while pet cats usually lived downstairs with the servants in the kitchen and out-houses – not the worst place for an animal that enjoys warmth and food! In contrast, stable cats and farm cats lived outdoors all the time and dogs for shooting were in kennels.

Feral cats, then and now, live wild throughout the world, find shelter wherever they can – city office buildings, old sheds, remote barns, factory premises, bushes and trees, woodpiles, disused animal burrows, sand dunes and even caves. Cats can adapt to all kinds of conditions.

Hurrah for the cat flap

The cat flap or kitty door is probably the single greatest invention in modern cat keeping. Its invention is often attributed to Isaac Newton (1643–1727), the great English physicist, mathematician, astronomer, natural philosopher, alchemist and theologian. He is said to have cut two holes in the door of his room – one large one for his cat and another small one for her kittens. A delightful story, even if it is probably untrue.

Cat flaps, or kitty doors, give a cat choice – and cats love choices. Give some thought about where to put the flap, remembering cat preferences not just human convenience. A cat flap into the garden is a safer choice than one straight onto the road (even a small road). Cat flaps are usually cut into doors but there are some available that can be put into windows. This means that even people who live in a flat on the first floor can put a cat flap into the glass of their windows and, if they provide a cat 'ladder' of wooden steps, the cat can have access to the garden.

Cat flaps can also be fitted to garages, garden sheds or stables, if you are providing shelter for a semi-wild cat or a feral cat. Nowadays there are some very ingenious cat flaps. Almost all the ones that are commonly sold can be locked both ways, or simply locked one way. That means, for instance, that the cat flap can be left open for incoming cats and shut for outgoing ones – or vice versa.

Better still, there is now a cat flap, PetPorte, that keys into your cat's microchips or even into the microchips of several cats. This means that your own cats can come and go and their home territory is secure, as neighbouring cats or hungry strays in search of a meal cannot intrude. Very few cats like having their home invaded, and this cat flap provides the perfect solution. Cat flaps of the PetPorte kind are also useful indoors for giving timid cats a refuge inside the home, where they can get away from toddlers, puppies or other felines bent on harassment. There are also magnetic cat flaps, worked by a magnet worn on the cat's collar. However, these may result in the cat collecting small items of cutlery on her collar. I know of a cat that got stuck by its collar magnet to a night storage heater!

Teaching your cat to use the flap

Many cats manage to teach themselves how to use a cat flap without any difficulty – particularly strays that are searching for a free meal. If you are getting a cat from an animal shelter, ask them if the cat knows how to use a cat flap. If they don't, even an elderly cat can learn – as long as you remain patient. Don't just stuff your cat through the cat flap; it is tempting to do so but the cat will be so filled with indignation and panic that she may refuse ever (and I mean *ever*) to go near the flap again.

Instead, stay patient. Clip a clothes peg (or similar) to the door of the flap. If you put it high up near the hinge it will hold the flap wide open. If the cat is indoors, fix the peg so that the cat flap is held wide open on the outside. As soon as the cat has gone through the now-open entrance, change the peg to the other side, to hold the door open on the inside. Do this for at least a week. It's maddening at times because you have to keep rushing to the cat flap to change the peg, but remind yourself that it does work.

The following week, fix the peg a little lower down so that the door is slightly less open. It should still be at least two-thirds open. This means the cat has to squeeze past it a little. The following week, keep it half open. If at any time the cat seems to be having difficulty, go back a week to a more open door. Then follows a week with the door a quarter open, and in the week following, open it just a tiny bit so that she has to push it with her head. Voila! The cat has learned to get through.

The occasional gift of prawns held the other side of the cat flap will help entice your cat through at the start of the learning process. I have even taught a thirteen-year-old cat to use a cat flap – though it took a lot of human patience and three months of using the clothes peg.

Keeping your outdoor cat safe

Regular vaccinations and flea treatments will protect your cat from some of the diseases and parasites to be found on other cats; microchipping will give you a better chance of getting her back if she is lost; and neutering is essential. Un-neutered tomcats are more likely to get into fights, catch diseases, leave home for days and spray in your house. Un-neutered females will disappear to find a mate (or several!) and leave home if they decide to have their kittens in a place outside which they consider more secluded.

About sixteen out of a hundred cats, according to one estimate, will be run over by a car and the number of cats killed every year on the road in the UK may be almost a

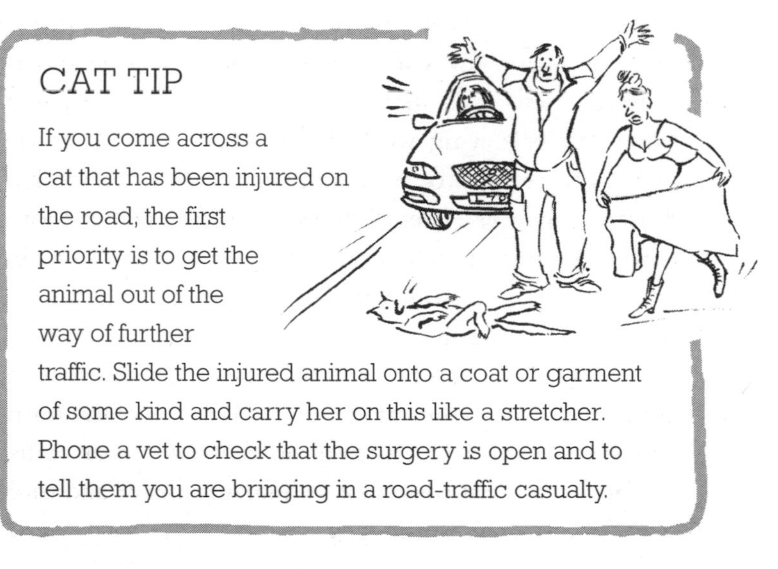

CAT TIP

If you come across a cat that has been injured on the road, the first priority is to get the animal out of the way of further traffic. Slide the injured animal onto a coat or garment of some kind and carry her on this like a stretcher. Phone a vet to check that the surgery is open and to tell them you are bringing in a road-traffic casualty.

million. Some people have put up warning signs on their own property, but in general nothing is done to remind car drivers of the danger of killing a cat.

The cats most at risk are, of course, those that are allowed to roam freely whenever they choose. Young cats, between the ages of seven months and two years, are more likely to die than older cats, and males, even neutered ones, are more at risk than females. This probably just reflects the fact that young males, even neutered ones, are more energetic and roam further and for longer.

The most dangerous time is at night, when drivers are unlikely to see a cat as she dashes across the road. They may not realize they have hit one and, even if they do stop, may not be able to see the victim as she drags herself away. As for the cat, although she can see much better in the dark than humans, she sees less well in bright light, so car headlights dazzle her. Also, cats just cannot judge speed and they will make a run for it at the worst possible time.

You can protect your cat from the danger of the road by keeping her indoors at night. So why don't more people do this? One reason is that some of us cat owners just hate denying our cats what they want. Some cats have never been thwarted in their desire to be out in the twilight hunting, yet if a new regime is started they will adapt, even if it takes two or three weeks for them to adjust to the new timetable. The older the cat, the less likely it is to go off on long-range hunting expeditions.

The need for a litter box may be another reason why people don't like keeping their cat indoors at night. But all homes ought to have a litter tray, partly for the benefit of a cat that may not wish to go out in the rain and snow just to urinate, but also for the benefit of neighbours. Indeed, not only should you supply a litter tray indoors for your cat, you should also supply a latrine outdoors in your garden, not somebody else's. Dog owners pick up dog poo; cat owners should be willing to take responsibility in the same way.

Cat mess in the garden is not appreciated by those who have cats, let alone those who don't. Just letting your cat defecate next door does not lead to good relationships with neighbours; it could also put your cat at risk from an enraged nearby gardener. In suburban areas where there are lots of cats, your cat may suffer great anxiety from having to go through enemy territory every time she needs to take a leak. A place to go in her own garden, as well as a litter tray inside, will help her feel more secure.

CAT TIP

Make a latrine for the cats in your garden. Dig a hole in a dry place – under a shed, under a hedge, under shrubs – and fill it with gravel. Over the gravel put a thick layer of dry sand or fine soil. Put some cat-soiled litter there to encourage use at the start. Scoop out the poop and wash urine away with a hose pipe regularly – remembering that the latrine will not be used until it has dried out.

There's an old wives' tradition of actually locking cats outside at night. This is dangerous for all cats and positively cruel to elderly ones. They have to shiver through cold nights instead of being inside in the warm, and they are also at greater risk of road-traffic accidents. People who insist on putting their cats out at night shouldn't have them in the first place.

CAT TALE: George, my beautiful sleek black cat, vanished one October night. He was nearly two years old and at the height of his hunting powers. In the previous 12 months he had hunted not just mice, but also rabbits and birds. Like a good cat owner, I locked the cat flap at night. He usually arrived home at 8 p.m. but on this particular evening failed to turn up. It was a moonlit night and I thought that perhaps he was just staying out late on an unusually long hunting expedition. There had been one other occasion when he had failed to come home until 3 a.m. Of course, I called, went out with a torch, but I wasn't that worried.

The next morning he was still not home. I walked every single road that touched on the land round our country house. I walked every single hedge of the nearby fields, calling as I went. I leafleted four villages, all the local vets, and the local animal shelters.

He disappeared forever. I never saw him again. I never even found his body. Perhaps he had been run over and crawled away into the undergrowth. Perhaps a fox (for there were plenty) had jumped him, unaware. These are the dangers of letting a cat roam free.

Teach your cat to come when called

In the Northern Hemisphere, it's not always easy to make sure your cat is indoors at the time you want, because there daylight hours change according to season. The solution is to teach your cat to come in when called. Why people don't do this more often, escapes me.

Cats are trainable. Admittedly, they are not as easily trained as dogs, since they have no great motivation to please their humans – 'I'm not bothered' is a cat's response to a human request. So give them a reason for bothering: food. Not just any food, though. Offer them extra-special food in the form of cat treats, small pieces of real meat, dried kitten food (which is higher in protein and fat than normal adult food), or pieces of chicken or prawns.

Decide on your call – the tone and sequence of it. Just enunciating the name may mean nothing to your cat, but a good call, on a familiar note or sequence of notes, will be much more easily heard. If you have more than one cat you should have more than one call. You might call one cat on a single repeated high note, and the other on a lower sequence of two notes, for instance. Be consistent.

And don't forget the food reward. Why would a cat put herself out for nothing more than your praise? They just don't care that much for praise. Pay them. Pay them well. Make it worth their while, and they will come home when called. Except, of course, if they are in the middle of a most exciting mouse hunt. In a cat's mind, the best treat in the world does not compare with the absolute joy of hunting.

The garden is full of corpses, can I stop my cat hunting?

You can't stop your cat hunting, and nor should you. Hunting is not just what cats do: it is what cats are. It is as natural for a cat to hunt as it is for us to love. And it is as joyful too. Yes, it is horrible to have tiny corpses, or even big ones like rabbits, strewn about your garden or brought into the house. And yes, it is even more horrible to see a cat playing with a mouse, but that is natural too. What we are seeing when a cat plays with her prey, is a cat displaying hard-wired behaviour, as I explained in Chapter One (see pp. 11–27). This is what zoologists call a fixed action pattern, i.e., a pattern of behaviour that is relatively inflexible. Once the behaviour has started, it is likely to continue to the end of the sequence.

In the case of a cat that is not hungry or has caught prey that it does not wish to eat, the hunting sequence gets shortened. Eye-stalk-pounce-grab-kill-eat becomes eye-stalk-grab. Rather surprisingly, if the mouse stays absolutely still, the cat may also (though not always) stop the play, for the hunting sequence needs to be triggered by the movement of the prey. So if the mouse stays still, the cat won't or can't continue the hunting sequence.

This is seen most easily when a cat is playing with a dead mouse. This kind of play is less painful for an owner to watch. It is also more relaxed than the play with a living mouse, in that the cat, as well as pouncing and grabbing, may roll on her back or leap in the air. You will also see your moggy biff the little corpse with her paws, moving it across the ground or even catching and throwing it in the air.

By doing this, she makes the dead mouse move. And the movement triggers off the hunting game all over again. (The same thing happens with cat toys; a motionless toy is not exciting, but a moving toy excites the cat to play.) Some cats will play for half an hour or more, occasionally pausing to get their breath before starting again.

We humans may have moral doubts about this play with prey, but anybody seeing a cat doing it knows that they are lit up with excitement and joy. Cats are not deliberately cruel. Yes, what they do is cruel, very cruel indeed for the poor little mouse, but this is not deliberate or thinking cruelty on the part of the cat. It is unthinking, involuntary behaviour, that has been hard-wired into the cat's brain by years of evolution in order to make sure the cat survives. Cats presented with prey do not choose to hunt; they are obliged to hunt.

Can you protect wildlife from your cat?

You should not stop your cat hunting, but you can at least make it easier for her prey to escape. Keeping your cat indoors at night will help save the lives of the birds that come down to feed on the ground at dawn and dusk, the favourite hunting times for cats. It will also reduce the hours that she hunts, thus depleting her game bag. This is probably the single most important and easiest way by which you can protect wildlife.

Cats catch mainly mice, shrews, young rabbits and other small mammals, but they will also catch birds, lizards and insects – more or less anything moving. Bird lovers find their hunting hard to bear, and in Australia, where the small marsupial mammals are already under threat, cats are a real menace to endangered species.

Cats will even hunt snakes. It is usually a long drawn-out affair with the cat poking at the snake then retreating, and to the observer it looks as if the cat is tiring out the snake before it dares go for the final kill bite. When snakes are poisonous, it is obviously a dangerous hunt, but, as I have said, cats don't weigh up the danger. When presented with a target, they have to hunt.

Some people put bells on their cat or use various collar devices like the cat bib. I myself don't favour these because I believe even the so-called quick-release collars often don't release themselves quickly or safely enough. Most cat-rescue organisations have come across stray and starving cats with one paw caught in their collars, which are biting into the flesh of the neck. Even so-called quick-release collars can do this to cats.

Wildlife casualties can also be reduced by careful management. Oddly enough, in gardens where the birds are fed at a bird table, one survey suggested that the casualty rate is lower rather than higher. Careful siting of bird boxes, bird tables and feeders to positions where a cat cannot lie in wait will help. So too will spikes to stop cats climbing trees, and scrumpled netting placed under shrubs from which the cat could ambush ground-feeding birds. Planting spikey shrubs such as holly and berberis in areas where birds need to perch will also deter cats.

CAT TIP

Cats often hunt and bring home frogs from nearby garden ponds. Stopping cats catching them is impossible, but giving the frogs plenty of hiding places will reduce the numbers killed. Piles of wood, heaps of big stones, plenty of pond plants and long grass around the pond will provide places where cats cannot get them.

CAT TALE: Clari, a ginger-and-white neutered tomcat, regularly hunts and kills snakes. He lives in the foothills of the Pyrenees in South-west France, where both adders and non-poisonous smooth snakes abound in the summer. His owners, Giles and Lisa, often find the mangled bodies on their patio on summer mornings.

Watching his method of preying on snakes is extremely worrying for his owners, who fear that he will get bitten. 'Once he catches sight of a snake, he goes on "all alert" and then, judging the moment very quickly, his first move is to seize the snake in his jaws at a point about its middle, so that it hangs down on each side of his mouth. He then runs fast to deposit it either inside the house (to the general panic of his humans!) or into his favourite thyme bushes on the terrace,' says Lisa.

If he has taken the snake into the house, and if Lisa and her husband Giles are around to see, then human intervention occurs, as the idea of it getting lost under the furniture and creeping out at night or whenever frightens them – as well it might. Often during this process (doors left wide open, rolled umbrellas and walking sticks featuring large), Clari again seizes the snake in its middle and this time runs to dump it in the thyme bush.

Once it is there, Clari can survey its every movement and it can't escape without him jumping upon it. He does indeed stir it into action by poking at it, if it lies still too long (too long for Clari). 'He may well be relying on tiring it out, but he is also I think waiting for the moment when it is positioned just right for the fatal blow,' says Lisa. This is the killing-bite in the neck area. So far, he has been too quick to get bitten by a snake.

Cats and garden fencing

Cats don't recognize human barriers such as fencing or walls; these are merely obstacles to be overcome and any self-respecting cat will roam far further than their owner's garden. Some ordinary suburban cats may have a territory as large as 17 hectares during the day, and at night they tend to roam even further. One surburban cat was fitted with a radiotelemetry collar, which revealed that she covered a territory of nearly 28 hectares. No wonder, therefore, that it is at night that cats are most likely to be run over. We often don't have the slightest idea what our cats are doing or how far they have wandered from their home, but it is a good guess that they will have travelled to a field or wood where good hunting is available.

If you want to keep your cat in your own garden, you will need very special fencing. It must be about 1.8 m (6 feet) high and be made of stout fencing materials, with a small mesh. A horizontal overhead section, pointing inwards at about 45 degrees, will prevent the cat from scrambling up and over. If you want to keep neighbouring cats out (and this may be important in an area densely populated with cats), you need the overhead section pointing outwards – only remember that, if you do this, your cat may be able to climb the fence but will be unable to return. If you are good with DIY, it may be possible to have both an inward-pointing and an outward-pointing section at the top of the same fence. Planning permission may be needed.

CAT FACT: Cats hunt where they are most likely to find suitable prey. Some studies have found that they will go straight to a freshly mown grassfield, or a recently harvested cornfield, or even a new clearing in a wood, because there are more mice there. On a remote island in the Seychelles, cats prey on baby turtles. When the turtles are hatching, tracks in the sand show that the cats will go there to hunt.

Garden dangers for cats

Even if you have fenced in your garden properly to keep out neighbouring dogs, there are other hazards within garden boundaries: water butts, for one. A completely open water butt is probably relatively safe, because, if a cat falls into it, she can pull herself out by her paws. Water butts with covers are more dangerous, though, if the cover is defective. A Siamese who disappeared was found several weeks later, drowned in the water butt. She had fallen in when the cover gave way in its centre, but the cover circumference was too strong for her to get out again. Swimming pools that have no way of exit for animals are another obvious hazard.

Other dangers are less obvious. A feral cat, Sam, was found wandering with a drain cover weighing 1.3 kg (2.8 pounds), a quarter of his body weight, stuck round his neck. He had to be anaesthetized by a vet so that the drain cover could be lubricated and eased off. Cats have also been drowned by getting tangled up in netting put over garden ponds to deter herons. Garden chemicals can poison cats, too. What usually happens is that the cat gets the chemical on her body or paws, and then licks it off, thus swallowing it.

The outdoor life for a cat has many pleasures and as many risks. The indoor life has far fewer pleasures and far fewer risks. It is up to you to decide which is best for your cat.

CAT FACT: Here are some of the things outdoor cats have raided from neighbour's houses and gardens:

- Claude stocked his owner's pond with goldfish and even a large Koi carp. Nobody ever discovered whose pond he was raiding.
- Domino brought home a large blob of frogspawn. 'About a mugful', estimated his owner.
- Thai-Moon, a Siamese, brought home a living and very pregnant white mouse from an aquarium (with a top to it) in a neighbour's shed.
- Ranji, another Siamese, leaped on to the kitchen table of the next-door house while the occupant was answering the doorbell. He stole a kipper from her plate and took it home to his owner. The neighbour followed shortly, asking: 'Where is my Sunday breakfast?'
- Timmy brought home a pair of underpants – male and clean. Each Sunday he would bring in a small Yorkshire pudding and, just a week before his death, he managed to hunt down and bring home a freshly baked fairy cake, still warm from the oven.
- Blot, a long-haired black cat, appeared in the garden dragging something behind him. With much pulling he reached the back door. He was dragging a 1.6 kg (3.5 pounds) frozen chicken in its plastic wrapper.

CHAPTER SEVEN

The Indoor Cat

'When I am playing with my cat, who knows whether she have more sport in dallying with me than I have in gaming with her. We entertain one another with mutual apish tricks. If I have my hour to begin or to refuse, so hath she hers.'

Michel Montaigne, 1533–92

Our clever cats have moved into the lap of luxury in the twenty-first century – or what we humans think is the lap of luxury. Many of them are kept inside the house all day and all night with warm-as-toast central heating in the winter and air conditioning in the summer. They don't have to mouse for a living. They are indulged and pampered like poor little rich kids, with nothing to do except eat, sleep and dream of the mice outside.

In some ways it is a cushy life: an indoor cat is protected from disease, from accidents and will often lead a longer and healthier life than a cat that is allowed out of doors. The American Association of Feline Practitioners is in favour of urban and suburban cats being kept indoors or, if they are allowed out of doors, being given an outdoor enclosure of some kind. Some pedigree cat breeders even refuse to sell their kittens unless they are going to be kept indoors.

Yet, is all this luxury and safety good for them?

Which cats make good indoor cats?

Feral cats, young and active cats that are used to outdoor life, or healthy strays that have lived successfully on the streets, will rarely make good indoor cats. If you are getting your cat from a rescue centre, ask about the cat's previous life before assuming that he will enjoy indoor life. Some won't. Others, particularly those that have been rescued from animal hoarders and have thus spent their lives so far indoors, will be fine having never known the joys of freedom outdoors.

Elderly cats, particularly those that have suffered from living rough on the streets, will often settle gratefully into an indoor environment, thankful for the peace and quiet. Even a cat that is used to outdoor life may settle happily indoors when he reaches old age. Blind or deaf cats usually need the safety of an indoor environment anyway, and FIV-positive cats are a safety risk to other cats and therefore should also be kept indoors to stop the spread of disease. So in almost any rescue shelter there will be cats available that are suitable for indoor living.

The safety of the home, however, may cost a cat some of his happiness. It is certainly asking a lot to take a cat that has been used to hunting in nearby fields and stables and expect him to settle down indoors with no chance to roam and hunt. Even cats that have never been allowed out of doors, and so have been used to the indoor life since kittenhood, may get restive indoors if there is not enough for them to do.

Nor is it safe to assume that pedigree cats will grow up more docile than moggies. Some pedigree breeds are particularly active and even Persian cats, traditionally thought of as lap cats rather than hunters, have strong hunting instincts. Like a child that is overindulged, under-exercised, and given no adventure, a bored cat may become an aggressive brat, a cat behaving badly because there is nothing much else for him to do.

CAT FACT: Cat behaviour counsellors report a higher than normal incidence of behaviour problems among indoor-only cats. This may be because indoor cats develop more problems, or because their owners notice them more.

How to keep an indoor cat happy

Any cat kept entirely indoors is like an animal in a zoo – a good zoo with expensive cat food and luxurious surroundings. Yet the indoor cat has lost the opportunity to exercise his hunting instinct to the full. His power of choice, so essential for a really happy lifestyle, is also drastically reduced. A cat with his own cat flap can choose whether he goes out to hunt in the garden or comes indoors to play a game with you. He can sit in the sunlight or lie in front of the fire. Outdoor cats are rarely bored, but indoor ones can be. And bored cats easily become bad cats.

So what can you do? You will have to make some changes...

The indoor cat is entirely dependent upon us humans, not just for food and shelter but for every single activity – pouncing, watching, stalking and playing. This dependence brings with it great human responsibility. Your cat needs your attention and, if he doesn't get it, he may demand it by ambushing you from behind doors as if you were a mouse, biting your ankles, demanding food with menaces, and tearing round your flat in the early hours of the morning making sleep impossible for you.

The secret of a happy indoor cat is keeping it busy. The buzz word for this is environmental enrichment. It means that you must make sure that your cat has opportunities to be active and do things that cats do. This is not as easy as it might seem; left to their own devices with nothing else to do, cats will fill their time with sleeping and eating. We humans are also naturally lazy and spend too much driving instead of walking. We get fat. So do our cats.

CAT TALE: Vincent is a pure white, deaf cat who lives with his owner, Pam. Most of his life he has been an indoor cat because of his disability. This has encouraged him to develop interesting little behavioural quirks which give much amusement to Pam. In particular, he spends a lot of time placing items in his food bowl. These are often soft toys and toy mice. However, at times he has been more inventive, placing gloves, small books, towels, saucers and once even a complete telephone handset! He even dragged an entire full-length coat over the bowl. He also tries to drag the kitchen mat over the bowl and will place stones in it too.

Vincent is a very intelligent cat; he worries when the water level of his bowl is low and uses his paw to check the level. If it is high enough, he drinks from the bowl; if the level is low, he dips his paw into it and drinks the water off his wet paw. He also enjoys the flicker of light and shadow on the walls.

He amuses himself by throwing things about, such as ornaments, clocks, and books. Sometimes he does this to get Pam's attention, and at other times he will do it just for his own fun while she is out. He once pushed over a row of jams and pickles. Unlike most cats, he enjoys playing with the vacuum cleaner and shows no fear of it – perhaps because he is deaf and cannot hear its noise.

Vincent is very creative in the way he uses cardboard boxes. If boxes are placed with the open side downwards, Vincent will wriggle underneath. 'It is disconcerting to see a box moving across the room, apparently by its own volition,' says Pam. He also enjoys jumping in and out of them. 'There are problems when Vincent jumps onto a box which he thought was open and finds that it isn't. Closed boxes are also splendid launching pads to reach glass, china and plants that seemed out of reach,' adds Pam.

Pam also plays a game with Vincent using yards and yards of brown paper. 'I allow Vincent to romp, bury, slither, hide, leap and rampage in the paper. And then he sleeps in it. When he grows bored with it, I put it back into its box, lie the box on its side and let him find it again.'

One or two cats – or just one more?

It may seem a good idea to get more than one indoor cat, to make sure they are company for each other, and if cats were completely social animals this would be the obvious thing to do. But cats are not always social, and if they fall out in the confined space of an indoor life, there is real trouble ahead. Unrelated cats that have a cat flap usually manage to keep out of each other's way. Cats that live entirely indoors can't do this so easily in a small house or flat.

If you decide on having two cats, the best way to do this is to get two kittens from the same litter. Related cats of the same age, brought up together since kittenhood, are more likely to groom each other, sleep together, and generally form a little family. You might also be able to adopt two cats that have already lived together from a rescue centre – but make sure you can take one back if there is trouble between them. Sometimes cats that get on in a rescue centre (united against a hostile world) start quarrelling when they get into a home. Or rescue centres assume that cats that came in together are friends – and they are not.

Don't be tempted to fill the house with cats. Animal hoarders who keep too many animals are putting their own selfish desires before the welfare of their pets. As a rough rule, two indoor cats are usually enough for a three-bedroom flat or house. The more unrelated cats you keep, particularly indoors, the greater the chance of ill-feeling between them. Cats need to be able to keep between one to three metres' (roughly three to nine feet) distance from each other at all times. This is a minimum not a maximum range, and a cat that is forced to live in close quarters with another cat he dislikes may fall ill with stress.

It is also important to be generous with litter trays: one litter tray per cat is the minimum and these trays should be placed in different locations, not side by side in one place. Be equally generous with cat beds – two per cat, preferably in four different locations, will allow each cat a choice. These need not be expensive – just cut down a cardboard box at one side so that a cat can see out, and add a fleece or an old piece of blanket.

Having just one feeding location, where both cats have to eat side by side or share a water bowl, is also unnatural for these solitary hunters. So it is better, if you have more than one cat, to have more than one feeding location and more than one water bowl. Better still, as I will explain later (see pp. 143–6), make feeding into a hunting game throughout the house.

CAT TALE: When Whee and Chainy first started living together, they got on very well. As brother cats, they played together and slept together with never a cross word, so to speak. They were company and entertainment for each other. It seemed an ideal loving relationship, until Whee had to go to the vet. The first fight happened when he returned from the vet and his brother Chainy attacked him, probably because Whee smelled wrong. It was a bad fight. Whee was very frightened and ran away and Chainy chased him. A second fight occurred the following day, then a third one. Beatrice, their owner, was immensely distressed.

She decided to see if enriching their personal space with more to do would help. She installed a huge floor-to-ceiling cat tree, bought lots of interactive toys and held at least two intensive play sessions every day. She gave both cats lots of attention and praise for good behaviour. She also plugged in a Feliway® diffuser and administered medication prescribed by her vet.

Despite all this, the fights got more and more violent. Chainy would launch completely unprovoked attacks, to the point that Whee, who had always been a very confident cat, was spending most of the time just sitting on his bed or walking around carefully. He jumped anxiously at every noise. Chainy had turned from a best friend into a mortal enemy. Neither cat seemed able to forgive and forget.

Beatrice separated them into different rooms and there was an almost immediate change in Whee – he was playful like a kitten and came for cuddles. Still anxious, he refused to go anywhere near the closed door of the bedroom where Chainy had been shut in. Chainy was relieved to live separately and seemed very happy as he got full and undivided attention when

Beatrice was with him. The separation, however, was just too difficult in a relatively small flat. After several months, and a lot of heartache, Beatrice decided she must rehome Chainy for the sake of Whee. Although she had to give up Chainy, she knows that he will be happier in a home where he is the only cat, just as Whee is happy now he lives alone with her.

Think leopard – use vertical, not horizontal, space

Cats are not very keen on wide-open spaces. The huge flat floor of an empty warehouse would not be their preferred living space. So the quantity of space in your apartment will not, if it is just bare open floor space, be of great advantage to them. They are much more keen on quality vertical space. The single most important thing you can do for an indoor cat is to make it some interesting vertical sitting places. In an ordinary house or flat, some vertical space can be found by way of windowsills, cat-friendly furniture such armchairs or tables, and the very top of high cupboards or shelves. Clear some space on high shelves and place furniture so your cat can reach the shelves. Place a cat bed on top of a high cupboard or on a windowsill. This is just a start.

Making your home into a cat jungle

Some websites sell special built-in furniture for indoor cats – stairs that can be put up along the wall of your living room and shelves to perch on. With their innate sense of superiority to mere humans, cats adore being able look down on their owners from a great height. Wall stairs and perches will enrich even a small flat, making it possible for a cat to have useful, rather than useless, territory. A floor is not a good place for a cats to retreat to; a vertical stairway allows them to run out of reach of humans when they choose to be stand-offish or just do not fancy being petted.

Those with DIY skills can make a positive jungle of stairs, shelves, perching points and ramps so that a cat can move up and down the walls without having to touch the ground! Some people bring huge tree trunks into their houses, and fix these so that their cat has its very own jungle. Not only do cats enjoy using trees just as much as they enjoy climbing trees in the garden, but they will also use them as scratching posts.

These high points need to be designed for two different purposes. There should be high places that are viewing points for a cat to look down from and see what is going on below, and there should also be high places where a nervous cat can hide away from anybody looking at him. The latter is particularly important in a home with other cats or pets, or children. In the wild cats run up trees to get away from danger: a good indoor home should have an equivalent hiding place to run to.

Bell ropes, the kind used by bellringers with padded ends, can be hung for cats to swing on. Ball ropes will need a pretty energetic cat, so install them for younger cats that can be encouraged to use them. Small children's plastic slides might also amuse younger cats – see if your local boot sale can supply one. Don't buy one specially – not all cats will use these!

In your cat jungle, don't forget the kitty grass. Pots of this are now sold widely in pet shops. Of, if you want to save money, grow your own by simply digging up a suitable clump of grass. Be wary of adding too many houseplants to the cat jungle – some of these, such as poinsettia and lily, are fatally poisonous. When somebody gives you a house plant, check that it is safe before putting it in the house.

Some people grow catnip for their cats, but if you have a very enthusiastic cat you may find that the plant is nipped off before it grows more than about 5 cm (2 inches). One way round this would be to grow the plant in a room to which the cat does not have access, and bring it out for intervals of enjoyment.

CAT TIP

Shake a tiny pinch of herbs and flavours used for cooking over a piece of paper and hide these round the house for your cat to investigate. A cat's sense of smell is much more powerful than ours, so don't use too much.

Activity centres or just another couch on which to sleep?

Pet stores and internet sites sell structures called 'activity centres' or 'gymnasiums'. These are the big versions of the straightforward scratching post, made with sisal stems and carpet-covered ledges or tunnels. In the advertisements they look delightful, usually showing several cats using them at the same time. They look cute, but ask yourself: how well will they work for your particular cat?

They will be useful as scratching posts, but, as a way of getting a cat to do more, they may be rather a disappointment. Kittens will enjoy climbing them, hanging by one paw from the top, and batting any little table-tennis balls attached, but a mature cat is more likely simply to climb on and go to sleep on their ledges. The expensive cat gymnasium has become merely another expensive cat couch.

One problem with many cat gyms is stability. Some of the cheaper freestanding ones are not stable enough to stay static when a large, fully grown adult cat leaps on to them. If possible, test-assemble them first before buying. The larger versions usually need fixing to the ceiling and perhaps the floor too. These are expensive items and, if you have a limited budget, it may be that spending the money on cat stairs and perches will be more worthwhile.

CAT TIP

Cardboard boxes are a good way to enrich your cat's indoor space.

- To make a hiding place, cut an entrance hole on the left-hand side of the cardboard box and then cut a 'window' higher up on the right-hand side. The cat can sit inside and look out from the window.
- Undo both ends of a cardboard box to make it into a tunnel, being careful to allow a big enough box so your cat can't get stuck. Have a look at this site, http://uk.youtube.com/watch?v=5lj4n7n2uag A really big box can be made into a play pen. Put toys and food dispensers inside it.
- For cats that enjoy rummaging through paper, add some paper. Lack of space makes the activities inside the box more challenging.

I scratch, therefore I am a cat

Unless you actually like the frilled edges and interesting vertical marks produced by cat claws on your armchairs, get several scratching posts. An indoor cat has to scratch somewhere in the house and, if there are not posts available, he will scratch your furniture. You will need more than one post; indeed, if you care for your furniture, you would do well to install something for your cat to scratch in any room he spends time in. There should be scratching posts near doors that lead outside the home, and also next to the resting or sleeping area, as these are locations that the cat will want to mark as part of his environment.

CAT TIP

Make your own cat-scratching device. You don't want to encourage your cat to scratch on carpet so, if your apartment is carpeted, don't use ordinary carpet. Instead buy seagrass, sisal, or hessian matting. Tack this up against the inside of a door, or use it to cover a large breeze-block or a wooden box (the kind that comes with expensive bottles of wine inside).

As well as the rope-covered or sisal vertical scratching posts, there are cardboard-type scratching posts that can be placed horizontally or at an angle to the floor. Cats enjoy these too but they are messy, as small pieces of cardboard will fly off when the cat is scratching. If you wish to indulge your indoor cat, put out both kinds of scratching devices – horizontal as well as vertical. This gives the cat a choice, and it also more closely resembles what a cat would do outside – that is, choose which surface to scratch on each occasion.

Make your home into your cat's hunting range

'Throw away the cat's food bowl,' advises Professor Peter Neville, founder partner of the Centre of Applied Pet Ethology. An indoor cat cannot go and hunt for mice, but he can be encouraged to hunt for his food. Just putting the food in a bowl makes it too easy for a cat that needs activity.

Start by putting the cat's regular food in several small bowls all over the house or flat – using similar containers will help the cat recognize the fact that these mark where the food is to be found. To begin with the bowls should be easily visible – in the bathroom, in the bedroom, on the landing or stairs, in the living room as well as the kitchen. In order to get his meals your cat has to move from room to room.

Next, make hunt-the-biscuits more difficult. Place food underneath the bed (if you have a bed where there is room to do this); inside a large cardboard box with a cat entrance hole cut into it; on the top of a high cupboard; or out of sight behind a door or a curtain. If you have installed high cat walkways, put morsels of food high up on these, encouraging the cat to go all the way up to look for it.

Once your cat realizes he has to search for his food, get rid of the bowls and just hide the dry food itself. This provides you with some exercise too. As a devoted cat owner, you probably fed your cat its breakfast before making your own. Now you will have to get up earlier in the morning so you can rush round the flat hiding biscuits before settling down for your own coffee and orange juice! Both of you will stay slimmer!

CAT FACT: Indoor cats are safe from road-traffic casualties but are at greater risk of obesity due to lack of hunting exercise. If their indoor conditions are stressful, or if they live with other cats, they are more likely to fall ill with cystitis (feline lower urinary tract disease).

Making your cat hunt for his meals

Put food into puzzle containers; traditionally these are small plastic balls with a hole in which the food is inserted. The cat has to nose the ball around, or push it with his paw, to get the food out. Some cats, particularly greedy ones, catch on fast. Others will need a bit of help.

Motivation is important, so, before even offering a food dispenser, start by feeding the same food every single day for a week without any treats at all, then weigh out just a third of the familiar food into the normal feeding places. Buy some different dry food that you know your cat likes. Because cats prefer a novel food, this is what will go into the puzzle ball in order to motivate your cat. You need to encourage your cat to use the puzzle feeder, initially by adding food that will be more tempting, but if you simply fill the ball with treats every day, you will end up with a well exercised but very fat cat. After all, the idea of making your cat work for his food is to make him fit not fat.

For cats that seem to miss the point, teach them. Make the inner tube of a toilet roll into a puzzle ball. At first make very large holes in the tube and block the ends up with sticky parcel tape. The idea is that the holes are so large (usually much larger than in a bought puzzle ball) that even the most indolent cat cannot fail to discover that food is falling out of the device. Then make the holes smaller. Once the cat has got the idea that food results from the manipulation of an object, change to the plastic puzzle ball.

Fun with food

Some cats develop a bizarre habit of tearing and eating cardboard. Do not use cardboard boxes or lavatory-roll food dispensers for these cats. For more normal cats, you can make food dispensers out of various cardboard items.

- Use Smartie tubes with holes in the side.

- Leave a biscuit at the bottom end of a cereal packet so the cat has to hook it out.
- Tape the top of a cereal packet closed and make a large circular hole on one of the bigger flat sides. Be careful that this hole is either bigger or smaller than the cat's head so the head doesn't get stuck. If a cat can get stuck, he will.

- Place cat biscuits in a printer cartridge packet and close it up.
- Hide food between the pages of newspapers. You can tape newspaper pages into hidey holes or tunnels.

- Put food into paper (never plastic) bags. Tie the bag at the top (use wide tape or ribbon rather than string so that your cat will not swallow it) and hang it from a doorknob so it has to be pulled down and torn open.

Don't forget fun with the water bowl

Some cats are fascinated by water and will sit by a dripping tap, as if a mouse might emerge in the next drip. The movement and play of light on the water amuses them and they will often prefer to drink from the tap rather than their water bowl. It would be ecologically irresponsible to suggest that you leave a tap dripping while you are out, just to amuse your cat, but there are now devices on sale rather like miniature garden ponds, which circulate water so that it flows down a little channel into a bowl. Cats that like a dripping tap may find these of interest.

Even if you don't have a water-responsive cat, nevertheless leave more than one water bowl in the house. Most bowls of water are left near the food bowl – not the natural place. A cat in the wild catches a mouse, takes it to a safe place to eat, and only then goes off in search of water. He does not dine and drink at the same location.

Outdoor cats will often leave the house and drink out of a muddy puddle or the garden pond. Indoor cats don't have this choice. Since cats enjoy being able to choose, make sure there are bowls of water in at least two different locations. Even this small choice will help an indoor cat feel happier, and making sure a cat drinks enough is important to protect him against cystitis. Don't leave the water standing there so long that it collects dust on its surface.

Play is an outlet for the hunting instinct

Hunting for food will give an indoor cat something to do while his owner is out, but it does not fulfil all of his hunting needs. Just to recap, the full hunting sequence is eye, stalk, pounce, grab, tear and eat. This instinctive pattern of behaviour is set off by movement. Hunting for food, left either in small hidden dishes or even in a food dispenser, cannot completely meet a cat's need to stalk and pounce.

So, in order to fulfil its natural predatory needs, an indoor cat must have the chance to play-hunt, and he needs you to play with him. In the wild a cat would hunt about ten mice a day and only catch one in three, so your cat needs 30 pounces a day. If you don't give your cat enough play, as mentioned in Chapter 5 (see pp. 96–109), you may find that your bored cat starts treating you as the prey: ambushing you, pouncing upon you with his claws out, and biting your ankles. It is no fun to be greeted at the end of a tiring day by a cat that treats you like a mouse.

CAT TALE: Rhett Butler, a beautiful Bengal kitten, was terrorising his owners. He played very roughly – biting hard and clawing. Sometimes he actually lined them up in his sights, stalked them and then pounced and bit. Butler particularly enjoyed targeting his humans when they were in bed, pouncing and biting their faces if they weren't fast enough to duck under the duvet. Their movements while trying to escape him simply egged him on. Squeals of pain merely excited him more.

Butler was an indoor kitten. He spent five days in a small London house and then the weekends at a country cottage where he was also kept entirely indoors. Geraldine, his owner, complained: 'He seems to be learning to be very loving too, but this rather vicious biting is a bit worrying.'

Geraldine and her partner had to change their behaviour. During any game, if Butler's teeth or claws were used on his humans, that game had to stop immediately. They were to stay silent with no exciting cries of pain. They were to remain immobile for a minute or so if he attacked, so there was no exciting movement either! Butler's attacks were to be ignored. Only polite games with no claws and no teeth were allowed, and these were played as much as possible. Butler learned quickly. When his humans refused to play rough games with him he stopped playing them too.

The main reason for his desire to hunt humans was that Butler just didn't have enough to do. He needed the chance to practise his hunting skills and without any alternative prey, he had to practise on his humans. If he was to remain an indoor cat, a programme of immediate environmental enrichment would have been required.

Luckily, Geraldine decided that in the long run Butler would be happier with an outdoor life. When he was allowed outside, Butler started stalking and pouncing on alternative and smaller prey, so there was no further need to hunt humans!

Toys are not enough to keep a cat busy on their own. I have tested various toys designed for solitary play, but most of them are ignored by cats unless there is a human present to make them move. Cats need movement to unlock their hunting sequence. Unlike dogs, cats very rarely use a toy to play with each other. You will sometimes see a group of dogs chasing each other for possession of a ball, but this doesn't happen with cats. Sometimes a cat will take away a toy from another one, but they won't play a game together with it. Cats hunt alone, so, while a cat may play with a human, they won't play with another adult cat. Usually one cat plays with the toy while the other looks on.

Indeed, some adult cats will not play on their own except when presented with a catnip mouse. Even play with a catnip toy usually lasts only 10 to 20 minutes. Cats, unlike human junkies, seem to know when to stop sniffing their recreational drugs and don't develop a continuous drug habit! Even catnip toys lose their appeal as their scent fades.

CAT TIP

To keep a catnip toy well scented, buy some loose dried catnip and place it in a large jar with a tight lid. After the toy has been enjoyed, and (if necessary) has dried out, put it back into the jar. It will thus retain its odour.

Hunting games for grown-up cats

Fishing-rod toys are probably the most successful of cat toys. The little soft-toy mouse, so often attached to the end of a string in a bought cat toy, is there to attract the humans who buy the toy, not the cats who are intended to play with it. Cats do not for a moment think that a stuffed toy mouse is a mouse; they are attracted by its movement not its shape. These are good for humans, too, as it is possible to sit watching TV while moving the rod so that your cat is chasing the string. Another, rather similar, toy is a wand with a bunch of feathers on the top. This gives you a little bit more control over where the 'prey' is, since it will always be at the end of the wand. On the other hand, the unpredictability of a fishing-rod string, and its wider scope, is part of its appeal.

Another way of giving your cat a moving target is to tie a piece of trailing string to your belt so that you can 'play' with your cat as you move round the house. You can also throw balls of paper or kitchen foil or the corks from wine bottles. Devoted owners often walk round the house with a pocketful of corks to throw so as to keep their cats occupied.

Bought toys for your cat

Good shop toys include small, light fur-covered mice. Many cats will enjoy chasing these little fur mice over a slippery kitchen floor. It's not a good idea to leave all the toys lying around as then they will lose their value. Have a drawer full of them and put out some different ones each day. Cats enjoy variety. Cats may also discover the pleasure of playing in the bath. My cat Fat Ada would whirl round the end without the taps, clearly enjoying bouncing off the sides. A ping-pong ball in the bath can also make a good toy.

Some cats are turned on by light. You can now buy toys, often shaped like balls, that light up when moved. If you have a cat that seems sensitive to light, these are a good buy. Some cats, however, will ignore them completely. A drawback is that these toys are quite heavy, due to the required battery, as are toys that are wound up by a string or by a key. I have not found any that my cats responded to, and I suspect they are sold because they appeal to humans, rather than cats. Besides, we all know the great cat-shopping truth: the more expensive the toy, the less your cat will want to play with it.

Cats can make a toy out of anything

I have seen cats play with small items like paper clips or pieces of plastic bottle tops. These are light enough to bat over a tiled or polished wooden floor, but anything with a sharp edge is potentially dangerous. Similar but safer toys might be a bit of large dry pasta or even a large dried broad bean. (Smaller beans or small pasta pieces are more likely to be swallowed whole and might just get stuck on the way down.) Cat-chosen toys should be safe ones.

Cats also master the art of games that upset their humans. Curtains and blinds, for instance, often have cords with a toggle on the end for opening or shutting them. These make fine noisy toys for an indoor cat. They can be patted against the window or the wall, making a satisfactory noise. The humans in the house notice this, and often shout, 'Stop it', or even come over and pick up the cat. Bingo. This is a game that gets human attention big time. What cat could resist this magnificent pay-off?

Other attention-seeking games played by indoor cats include knocking ornaments off the mantelpiece, pulling down the trash can (with the extra hope of something nice inside), playing with a dripping tap, using the bath as a fair-ground wall of death by tail-chasing round its sides, pulling down tablecloths, and swinging from curtains. All of these get human attention very easily. If you don't want your cat to enjoy these games, you must learn to walk away from him rather than rebuke him. Like small toddlers, cats enjoy winding up their humans.

Let your cat eye the prey

Cats that can't hunt for themselves can at least watch their prey. This is the first stage in the hunting sequence of eye, stalk, pounce, grab, tear and eat. Make sure your cat can look out of the window by placing an armchair or a cat gymnasium in a position for him to do so. You can buy special cat window screens, or just make your own with wire mesh so that the windows can let in fresh air.

Put a bird feeder up in the garden for your cat to watch, or buy a bird feeder that sticks to the glass in a window. If this isn't possible, simply place bird food on the outside windowsill. Garden birds are quick to catch onto the chance of any free food, and even in cities there are pigeons who will perch there to eat it. Plant bushes, below the window, of the kind to attract butterflies. These fluttering insects make for good feline watching.

There is also a special bird box with an inbuilt camera so that the occupants can be monitored from a TV set. Most cats enjoy this greatly, even if they are not normal TV watchers. Cats love watching birds and by doing so they are fulfilling the first stage of the hunting sequence – and even better, the birds do not suffer as they do at the claws of the outdoor cat.

Less exciting to a cat but still worth watching are human passers by, dogs, wildlife and other cats. Indoor cats spend a lot of time peering from the windowsill if they get the chance to do so. If you don't have windowsills, arrange the furniture so your cat can watch.

Giving the indoor cat an outside chance

If you are lucky enough to have a closed patio attached to your flat, you can make use of vertical space in much the same way outdoors as indoors. This means attaching stairs, walk-ways, perches and boxes against the wall. These can be made more attractive by adding a large-scale trellis in front of them so that the cat walkways are half hidden. Growing plants up or against the trellis will give it a jungle look.

Be warned: if there is earth available in plant containers or pots, cats will use it as a litter tray. The easiest way to stop them doing this is to cover the earth with large rounded pebbles of the kind found on stony beaches. It is also important to make sure you are not planting plants that are toxic to cats either in your patio or indoors in your flat. When you are planning your patio garden, do remember that cats climb more easily upwards than downwards. If you place your walkways too high, your cat may climb up and over the wall, and then you have a missing cat either wandering round the rooftops or in a nearby road, unable to find his way home.

Roof gardens and balconies also offer space for cats, but these are potentially dangerous. Every year cats that have fallen several storeys are rushed to veterinary hospitals after they have lost their balance and fallen off windowsills or balconies, or from roof gardens. So these high areas need to be completely enclosed, like zoo cages, rather than surrounded by mere fencing. There are always escape artists in the feline world that will climb impossible fences. Cat websites can give several good examples of enclosures you can attach to windows, doors, or even cat flaps. Ordinary windows can be made safe simply by putting up screens, of the kind used to keep out insects, which can be attached to the outside of the window.

Walkies on a lead

Some outgoing and confident cats seem able to adapt to a harness and lead if these are introduced from a very early age, although most adult cats will have difficulty accepting them. Harnesses are often suggested as a way of giving an indoor cat outdoor experiences, thus enriching his life. My own opinion is that this idea is to be approached very cautiously indeed. Cats have the ability to wriggle out of a harness if they are thoroughly frightened – even if the harness is well fitted and has previously seemed perfectly safe.

Pepsi, a Siamese, was completely used to her harness and her owners were in the habit of using it frequently. It seemed as if nothing could go wrong. They took her with them by car all the way from Britain to Italy. It was on the busy Brindisi docks that disaster struck. They took Pepsi out of the car on her lead and she panicked at the noise of a lorry backfiring, wriggled out of her harness, ran off and disappeared forever.

Even if your cat wriggled out of her harness in a local park, she might well get lost. Frightened cats can run a long way and it is sometimes extremely difficult to persuade them to come back.

Grooming enriches your cat's life

Regular grooming, essential for all cat care, is even more important for indoor cats. Outdoor cats that hunt vigorously will lose some of their hair during their normal activities. Indoor cats, with less to do, will shed all their hair in your house and, because none of it is being rubbed off by foliage, they will probably feel the need to groom more. The result, familiar to all cat owners, will be hair balls, which have the wonderful scientific name of trichobezoars.

Hairballs are bad for your cat; they interfere with digestion and, if not sicked up on your bed, also cause constipation. All indoor cats, even the ones that have short hair, should be groomed regularly – once weekly for very short-haired cats and daily for long-haired ones. It is a very good way of bonding with your cat and giving it a pleasant experience. Regular grooming will also reduce the amount of hair shed on your carpets.

Clicker training your cat

Yes, you can do this. Clicker training uses a small device that makes a click noise and is the easiest way to train cats. You need a good food reward, as cats won't bother to do anything for nothing. Shrimps or prawns are usually highly motivating. Keep training sessions short, because cats lose interest very quickly. Two or three minutes at a time will be enough to begin with.

I have taught my cat William to shake paws, beg, die for his country, jump over obstacles and roll over. He doesn't do them perfectly, but then I am not a very good trainer, but he enjoys doing it and so do I. As soon as I show him the clicker he starts purring.

Start by getting a book on how to clicker-train your cat, as you need to understand the process, or ask a local dog trainer if you can attend her clicker classes to train yourself before going home to train your cat. Most dog trainers will be helpful, even though they may find the idea of cat training highly amusing.

CHAPTER EIGHT

Fat Cats and Faddy Cats

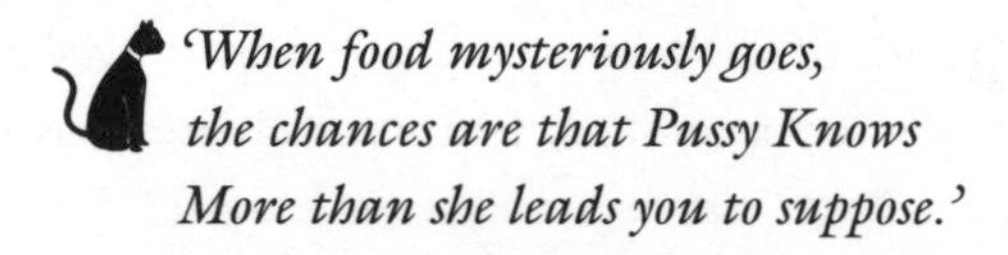

'When food mysteriously goes,
the chances are that Pussy Knows
More than she leads you to suppose.'

Anonymous

As your cat tucks into her cat food, you may well wonder why she insists on the most expensive brand. The reason lies right back in history when the feline evolutionary ancestors split from the other carnivores, most of which ate meat but also fruit and vegetables.

Cats, including the big cats such as lions and tigers, are 'obligate' carnivores, meaning that they are designed to eat other living beings. They will neither flourish nor survive on a vegetarian diet, and guess what? It is usually the most expensive kinds of cat food that have the most meat in them!

Feline teeth tell the story. Cats have fewer teeth than dogs and don't go in for prolonged chewing. They have big canine teeth to deliver a killing bite by crunching through the backbone of the mouse and further back in their mouth from the canine teeth are the carnassial teeth. These are slicing, rather than grinding, teeth, with flat sharp edges to slice meat off the bone.

Cats don't taste sweetness. Those cats that enjoy ice cream are not eating it for its sugar content but for the dairy fats that it contains. What attracts cats is fat and protein, just the things they would find in a diet of mice! Indeed, they can taste the difference between fresh meat and meat that is beginning to go off, which is why they refuse to eat carrion (as dogs will). Temperature, texture and smell also influence them.

Of course, a diet of small rodents freshly killed would be perfect for cats, but a diet of pure flesh meat isn't good for them. When eating the whole mouse, cats get fur, fibre, bones and gut contents as well as flesh – which is why many home-made meat-only diets don't contain the right ingredients.

CAT FACT: Feral cats are seldom fat, and most pet cats seem to manage their weight well. One cat in ten is overweight for no good reason. Scientists who study fat cats often disagree on exactly why they get fat. Factors within the home which predispose to weight gain include: living indoors only, sleeping a lot, doing very little when awake, being fed fresh fish or fresh meat, being fed particularly tasty food, having food put down all day, or having doting owners. Fat cats are more likely to suffer from diabetes, arthritis and other health problems.

My feline mum made me into a faddy eater!

Cats' food preferences are influenced by those of their mother. When scientists trained a cat to eat a diet containing banana, her kittens willingly ate banana too. So, if the breeder was feeding only one kind of food, as soon as you take your kitten home, feed her several different kinds of food, to make sure that she will be flexible about her diet. If you don't, as soon as the kitten joins the human household she will start making her food preferences known and you may find she will only eat one brand! And, yes, you are right, her favourite food will probably be an expensive one. Even if you do have a cat that will only eat a less expensive brand, you will still never be able to take full advantage of supermarket best-buy offers if your cat isn't flexible.

The other factor that influences feline preferences is monotony. When they are under stress, some cats may prefer a familiar food, but in normal circumstances most cats get bored with eating the same food day after day. They will eat it, of course, if nothing better is offered, but a new flavour stimulates their appetite. A new flavour is a better food to them, which is why there are rows and rows of different flavours on the supermarket shelves. Mind you, if you feed the new food long enough and stop feeding the old food, later on they will prefer the 'old' food when it is offered again – anything to break the monotony.

Cats like eating little and often – nibble eating. Cats in the wild don't eat huge meals: by catching small mice and birds they eat a series of small meals. If you leave food down for a cat to eat, they will have about eight to sixteen snacks a day, rather than two or three large meals. This is perhaps why cats seem better at naturally controlling their weight than dogs, who will eat two days' food in one session if they can get it. Nibble eating also seems to protect cats from developing bladder stones. Most cats that have plenty to do in their lives will not overeat, unless their owners continuously tempt them with extra food or treats.

Finally, cats have an inbuilt safety sense about food. If they get sick from eating a certain type of food, they will often refuse to touch it ever again. My elderly cat, Little Mog, was put on a special diet, but coincidentally fell ill with food poisoning, probably from eating a dying bird. Nevertheless, she steadfastly refused ever to touch that particular diet again. In her mind the diet, not the bird, had poisoned her!

How cats train their humans to buy the right food

How is it that most cat lovers end up feeding the most expensive food and, usually, feeding special little treats too? I have never forgotten going shopping with Jenny, a cat lover who was on a very limited budget. She bought a slice of nice steak and I commented, 'That will taste good.' In shocked tones, she replied, 'Oh, it's not for me. It's for the cats.' Her cats were eating better than she was.

You have to admire the amazing way cats train their humans to get their preferred (and mostly the most expensive) diet. It's quite an achievement. They can't fill their own shopping baskets, they can't even accompany their owners to the supermarket and, like toddlers, whine until the right packet is put in the trolley, yet still they manage to get their way.

The method is reward and punishment. We humans love seeing our cats expressing happiness. A cat that really enjoys her food goes to the bowl with enthusiasm, purrs while eating, and may even lick the bowl when she has finished. A cat that is not particularly enjoying her food will go to the bowl, then look up at the owner with a disgusted look on her face. Sometimes she will back away. At other times she will eat a little bit then leave the rest. Some cats will then claw over the bowl, as if they were covering up in the litter tray. Zoologists often say that they are 'cache-ing' their food, in the same way that a big cat will half-bury its prey so that it can come back and eat it later. I always feel (unscientifically no doubt), when my cat 'buries' his food in his bowl, that he is expressing his opinion along the lines that 'this food is s—t!'.

Either way, we cat lovers feel rewarded and happy when our cat likes her grub, and we feel guilty when she doesn't. We are co-dependent with our cats, I fear. Food is one of the ways by which we humans express our love.

Enabling the faddy and the fatty

Co-dependency occurs when a person focuses too much on the needs of another. They feel over-responsible for another's actions and put that person's well-being before their own. They put up with bad behaviour and make excuses for it. A cat lover, for instance, may start to 'enable' her cat's overeating, supplying gigantic meals on demand even though they know this is causing obesity.

Does this sound familiar? As the owner of a formerly fat cat, I recognize myself in this. Fat Mog (or Little Mog as I used to call her when I was pretending she wasn't really overweight) was just on the borderline of obese. Luckily she was a cat who could go out of doors and enjoyed hunting, so she never became really huge, but I am ashamed to admit that her extra kilos were my fault.

Once a year I would take her to the vet, and we would discuss putting her on a diet. I would come home with a large packet of diet food. I would swear that this was all I was going to give her. But within days, at the sight of her little black face pleading with me, I would weaken and 'enable' her overeating by giving her just a little bit – 'it can't do that much harm' – of extra food.

The same thing happens with faddy cats. There are cats who never finish their cat food because they will only eat the first half of the tin or envelope, while the rest has to go into the dustbin; there are cats that 'demand' a new and different flavour almost every meal; and there are cats who do themselves serious harm by refusing to eat anything but liver.

CAT TALE: Herbie, a six-year-old pedigree British Blue, is twice as fat as he should be. He is an indoor cat through choice, being frightened of going outside. This may be partly because he lives in a residential road where there are many other pet cats. He is also fearful of most humans, except for his immediate family.

Herbie lives for food. 'I have never seen a food bowl with food left in it,' says Jess, his devoted owner. 'He eats all the food, no matter how much is put in the bowl. He is not fussy; he will eat anything. If I put it down, he eats it immediately.' Herbie is not energetic; he spends a lot of time sitting, not doing anything much. He sleeps on the bed at night – though his huge size and heavy weight makes this difficult for both his owners.

He is given diet food by his conscientious vet which is weighed out daily and in theory he is meant not to eat anything else. 'We never give him titbits,' says Jess. But when I visited the household, with my own eyes I saw Jess take two slices of ham from the fridge and give them to Herbie. Somehow Jess cannot resist Herbie's pleading looks.

Much more exercise and more time spent hunting would help Herbie's weight problem. A programme of hiding food throughout the house might well mean he had to do more – if only going up and down stairs to find it. But because Herbie's owners cannot bear to refuse him food, it seems as if his weight problem will remain for the foreseeable future.

Confusing love with food

We humans are the ones that enable this behaviour. A very fat cat is not a happy cat. A cat that eats nothing but liver will suffer bone deformities and may die of vitamin A poisoning. A very faddy cat that insists on a new flavour or a new tin every single meal costs her owner a fortune.

Why are we humans such a soft touch when it comes to cat food? Well, for humans food is more than just food: it is an expression of love. We share our food with others round the table – unlike cats. We give food such as chocolates as presents – unlike cats. We woo people over candlelit dinners – unlike cats. Almost all social life revolves around food: cocktail nibbles at the annual office party, harvest festival suppers, Christmas lunches, club dinners. Food for us (unlike cats) is bound up with social life.

Cats cotton on fast to our weakness. They know that, if they want our attention, they can get it by asking for food. They will come and rub on us with an attractive purr. They may walk towards the bowl, sit down, look up and miaow. Or they do the tail-up greeting sign with a little tail flick. This is cupboard love, we realize, but we fall for it. With a truly aloof cat, the sort that just comes in for her meals, cupboard love may be the only love she gives us!

It is also easier to feed than to deal with the persistent interruption. How many times have you just pushed away your cat because you didn't have time to pet it? Yet, when she comes to you apparently seeking food, you have stopped what you were doing and checked the food bowl. Giving a bit more food may take less time than stroking her, meaning you can get back to work quicker. It is the easier option.

Some cats not only use food as a way of getting our attention, they start building in little rituals. They will take you to an already full bowl and apparently insist you should add food to it. They will stand refusing to eat until they have been stroked. They will only eat if you are standing near them. They will take food out of the bowl, put it on the ground and then eat it.

It's how, as well as what, you feed your cat

If you want a normal-sized healthy cat, not a large cat with a swinging undercarriage, start as you mean to go on. A cat needs to eat about the amount of food that will be used up in activity. It should equal out. What isn't used up in energy is stored as fat, resulting in what I call saggy tummy syndrome.

Choose a pet food that is high in protein, not in carbohydrate, since cats don't need carbs and carbs help make them fat. If you have a cat that's greedy, don't leave food down all day, and if your cat comes to you for attention, give her love and games, not food.

Where you leave the food matters too; you don't want every stray cat in the neighbourhood popping in for a snack. So don't place the food bowl just next to the cat flap, if you have one. Don't put it too close to the litter box, either, or your cat may refuse to use the litter box. Cats do not toilet in the same place where they dine any more than we do.

If you have more than one cat, more than one feeding station will be appreciated. We've all seen the photograph of a row of cats feeding close to each other from a row of bowls. It looks cute, but it's not how cats like to eat. They eat (and hunt) alone, not in company. Where food is left down all the time, I have rarely seen cats eating out of the same bowl. They usually time their visits to the food bowl so that other cats are

CAT TIP

If your cat is on a prescribed slimming diet, you can still feed her treats! Ask your vet to prescribe not just one slimming diet, but a second one, of the same slimming kind but from a different manufacturer. Feed the first food all the time but keep the second food to be used only rarely in small quantities as treats. The monotony factor will mean that these, though they are still strictly diet food, will be much more enticing. To make sure you don't end up feeding too much, calculate 95 percent of the main diet from your vet's instructions or the instructions on the packet, and measure this out for the day. Then calculate five percent of the 'treat' diet food and put that aside in a separate container to be fed during the day.

not around and they only eat with others when they will lose out by not eating immediately.

Finally, if you want to keep your cat healthy and slim, don't feed too many titbits. Of course cats would love to live on a diet of raw steak or cooked chicken, but this is not a balanced diet. Nor is pizza, curry, sardines on toast, ice cream, butter from the butter dish, cream, or asparagus tips dipped in butter. What kind of owner feeds her cat so shamelessly? Well, I have done this sometimes in the past but now succumb only very rarely.

How to slim down your cat – fun instead of food

Slimming down your fat cat is really difficult emotionally for most owners. It seems cruel to deprive your loved feline of what is most important in her life, i.e.: food. You must remember, however, that your cat will have a happier longer life if you can manage to do this. First stop is the vet for a prescription slimming diet, with clear instructions on how much of the food to give them daily. Ask for a regular programme of weighing, just like Weight Watchers for humans.

If your cat is truly enormous, it may be a good idea to ask the vet to take your cat into hospital to supervise the first week on the new diet. This will ensure that your cat diets under veterinary supervision day and night, and will start off the new regime sparing your feelings. After the first weight loss is measured, it may help you to stick to the diet regime back at home.

It's simple but difficult to do! Weigh or measure out the right amount at the start of the day, put it in a jar, and feed it throughout the day. Do not give your cat anything else – nothing, not the slightest bit of chicken, not a single treat. Nothing extra means nothing. Zilch. Not a fragment.

You can make your cat's meals last longer by hiding the food round the house. Make it easy for her to find to begin with, then make it more and more difficult. Use puzzle feeders so that she has to work for every single biscuit (see instructions on pp. 144–6), but remember that the food inside the puzzle feeders must not exceed the daily allowance.

The previous chapter also tells you how to enrich your cat's life with games. So, chase her up and down stairs, use a fishing-rod toy to give her exercise while you watch TV and, when she asks for food, throw a wine cork instead for her to chase. Fat cats are quite difficult to tempt into playing, but the more you do with her, the more she will respond.

Remember, you are going to replace food with fun. This is not deprivation. This is life enrichment.

Cats that sit in sinks or peer down the plug hole

Cats do very odd things with water, particularly those indoor cats with not much to do. They enjoy playing with water, pawing the surface perhaps to try to 'catch' the reflection on the surface. Others sit by dripping taps, as if they were mouse holes; they are apparently transfixed by the globules of liquid coming out of the mouth of the tap. They leap into sinks and baths, spending happy hours crouching in wait for something, perhaps a spider, to come up the plughole. Still others rush to the lavatory when it is being flushed and stare down the bowl to watch the water swirl round down into the waste pipe.

CAT FACT: Cats enjoy sitting in sinks. There is now a whole website devoted to this feline peculiarity: www.catsinsinks.com It shows cats asleep in sinks, kittens playing in sinks, fat cats squeezing themselves into a sink and just lots and lots of ordinary moggies sitting there.

Some cats persuade their humans to cooperate in weird drinking rituals. Lucy, a much-loved tortoiseshell, would only drink water from a glass or mug. At bedtime she miaowed until a glass of water was placed on the bedside cabinet and held by her owner while she drank!

Fresh clean water is an essential for every cat. Indeed, cats on dry food need more than one bowl because many cats dislike the human habit of placing a water bowl next to a food bowl. They will drink from it, because they have to, but if they are given a second bowl, they will drink more often. I have at least three water bowls for William, my cat – one near the food, one on his favourite windowsill and one outside the back door. He also drinks from puddles, my bedside water glass, the kitchen tap, and the bird bath. As he has a history of cystitis, it is important that he drinks a lot. His preference for puddle water probably occurs because he can taste the chemicals in tap water.

CAT TIP

Cats that have suffered from bladder stones or cystitis need to drink as much as possible to flush out their kidneys. Here are ways to encourage your cat to drink more:

- Feed tinned food, not dry.
- Add a tablespoon of hot water to the tinned food, making it into thick soup and thus simultaneously increasing its temperature slightly to make it more palatable.
- Place water bowls in several locations – in the bedroom, on the windowsill, near the food bowl and outside the back door.

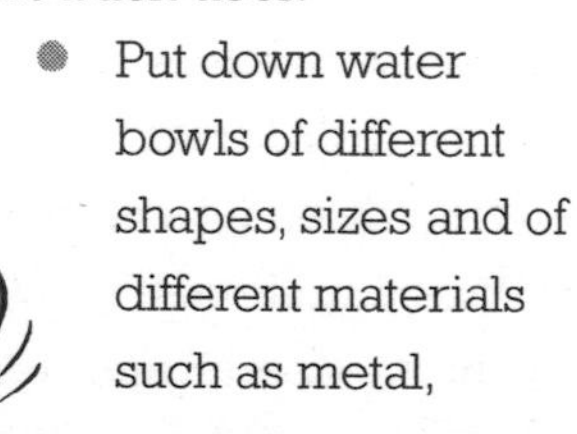

- Put down water bowls of different shapes, sizes and of different materials such as metal,

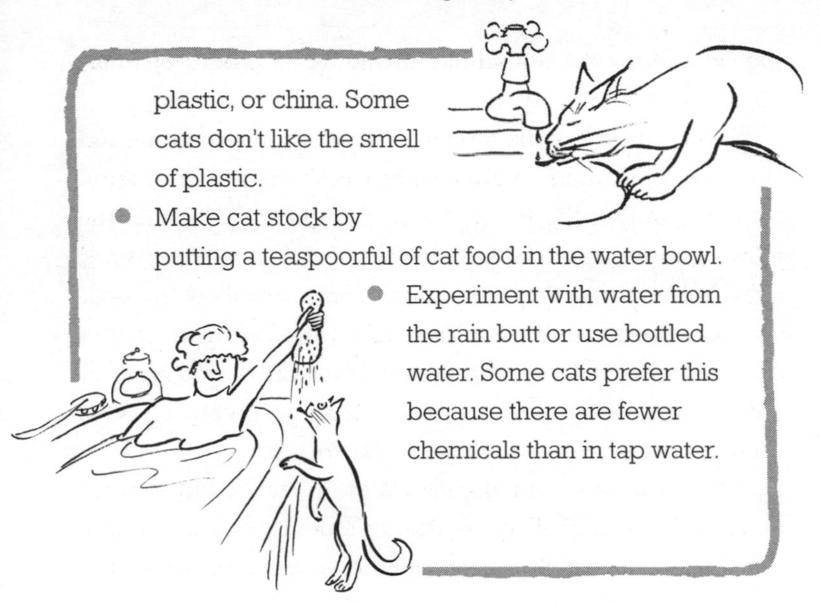

plastic, or china. Some cats don't like the smell of plastic.

- Make cat stock by putting a teaspoonful of cat food in the water bowl.
- Experiment with water from the rain butt or use bottled water. Some cats prefer this because there are fewer chemicals than in tap water.

Food allergy – what are the signs?

If your cat has digestive trouble, wind and loose stools, the first stage is a visit to the vet to rule out serious illness. Some cats with shaky digestions will benefit from a change of diet. As well as a diet for sensitive tummies sold over the counter, vets can supply all-in-one sensitivity prescription diets in both wet and dry forms.

Bald patches are often the first sign of food intolerance in a cat. The fur round their stomachs, their back legs, or the side of their bodies seems to disappear while the cat literally grooms itself bald. However, the most common reason for this is an allergy to fleabites, so proper flea treatment both for the cat and for the household, including all other pets, may put an end to it.

If that doesn't work, food intolerance should be suspected and a vet consulted. If after eight weeks on an elimination prescription diet the fur starts returning, it may be that the itchy skin was caused by a food allergy that made the cat feel itchy. Just putting a cat on an organic, a so-called 'natural' diet or an additive-free diet is usually not enough. Cats can develop a food allergy to different kinds of natural meat or cereals, not just to food additives. A proper hypoallergenic prescription diet from a vet is the way forward.

Overgrooming can also be a sign of stress, but it is far more likely to be an allergy to fleas or food, or even a response to pain. So check out these possibilities first before consulting a cat behaviour counsellor.

Cats that eat weird things

Cats sometimes have a taste for very odd food items. I know of cats with a passion for malt extract, olives, chilli con carne, chapattis, beetroot, malted cereals, tomato ketchup, Stilton cheese, marshmallows, toast with butter and marmalade, rice pudding, garlic sausage, chips, broccoli, melon and stewed pears. None of these items are good for cats, and in large quantities some could be positively harmful. So, if your cat has a taste for odd food, don't encourage it. A tiny bit, very occasionally, of what she fancies will do no harm; a large amount

CAT TIP

If a cat stops eating, take her to the vet. Loss of appetite can be a sign of pain or illness and must never be ignored. There are also various homely ways of encouraging appetite:

- Warm up tinned food so that it is at blood temperature.
- Sprinkle the food with an appetite enhancer. If you can't find one, try using a little goldfish food or a little ground-up malted cereal, such as grape nuts.

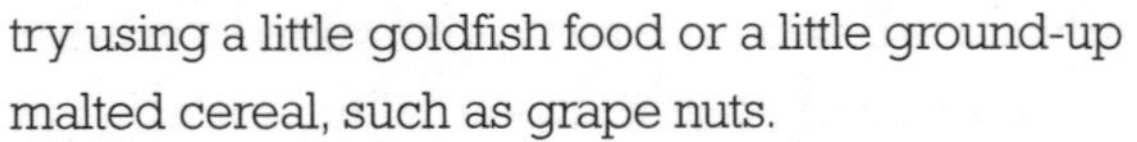

- Offer a new flavour or a new brand.
- Feed food with a strong odour. Mashed sardines, tuna or pilchards in tomato can sometimes kick-start a cat eating. Or mix just a tiny bit of sardine or tuna into the prescription diet.
- Hand-feed. I once started my cat eating by feeding her three little bits of kibble every 20 minutes throughout the day. It was time consuming, but it worked.

might. Also, the more you laugh or take notice of her eating, the more she will do it. Cats love to grab our attention.

More worrying are cats that chew wires. This can be an attention-seeking device. When my own cat, William, started chewing telephone wires on my desk, I naturally picked him up and put him on the ground. This, of course, was what he wanted. He had successfully interrupted my work on the computer and got my attention. The more I responded, the more he tried out this dangerous behaviour. The cure was, when I noticed him chewing wires, to say nothing, withdraw my gaze and walk out of the room. It took nerve to do this, but I persisted. When I did this consistently for a fortnight, he stopped chewing the wires. His frightening behaviour was no longer rewarded by my attention so there was no point continuing.

There are also cats that chew, but do not swallow, odd items like cardboard boxes or traditional basket-ware. 'Pica' is the name given to this behaviour. The normal human reaction to this is to point and laugh. So this too may be or may become an attention-seeking device and should be treated by withdrawing all human attention and leaving the room every single time the cat does this.

However, if withdrawing attention does not work, this may be a more serious form of eating disorder. Siamese, Burmese and oriental cats are prone to wool eating. They will chew and swallow wool, cotton, artificial fibres, and occasionally rubber or cardboard. At one point this was thought to be the result of not feeding enough fibre, so cat experts recommended feeding strips of newspaper with the food to fill the cat's stomach. Occasionally this worked, but it now looks as if something else is going on, something to do with the hunting sequence of eye, stalk, pounce, grab, tear skin or feathers, and eat (see pp. 19–20, 148).

Professor Peter Neville, founder co-partner of the Centre of Applied Pet Ethology, now believes that wool eating is a disorder of the hunting sequence. 'In the wild cats are capable of swallowing and passing feathers and bones in their gut, but they also have a behavioural requirement to skin or pluck their prey as well,' he explains. 'This involves pinning down the dead victim and plucking off the feathers or skin, and eating.'

Tinned cat food or quickly eaten dry food does not fill the stomach like a whole mouse or bird, which is why feeding bulky food may help stop the behaviour. Ordinary cat food also deprives the cat of the chance to tear and pluck at skin and feathers. So some cats, especially those that live without any hunting opportunities, look round for substitute things to pluck and tear and turn to cotton or wool. Siamese and breeds with Siamese blood in them seem particularly prone to this behaviour. Giving them the chance to tear off feathers or skin, by providing them with raw mice or chicks sold for reptiles, will often cure this bizarre behaviour, as the story of Garbo shows.

CAT TALE: Garbo, an eight-month-old oriental cat, started chewing holes in linen then moved on to chewing all kinds of material. 'At times the house looks as if it has been attacked by giant moths,' said her owner, Derek. He was mystified by this behaviour. He tried no-chew sprays, with no effect at all – Garbo just started on something else, such as towels or curtains. Her habit ruined several furniture throws, two complete cat beds and a small army of beanie babies. Garbo was an indoor cat sharing the home with other cats, but there seemed no adverse stress in her life.

Professor Peter Neville, founder co-partner of the Centre of Applied Pet Ethology, had the answer, which I passed on to Derek. He was to feed Garbo daily one raw turkey poult with all its feathers on. The poults were bought frozen from a pet shop that sold them for reptiles and Derek defrosted and fed one each day in the bathroom, away from the other cats, before going off to work.

This chance to pluck, tear and eat like a wild cat seemed to fulfil Garbo's needs. For the rest of the day, if she was still hungry, she ate the same food as the other cats. The mess (and it was messy) was contained in the bathroom. Derek would clean up the feathers and blood smears off the vinyl floor (being careful to wash his hands to prevent any risk of salmonella poisoning that comes with feeding raw meat).

Derek no longer had to hide away the towels, dishcloths and sheets! The giant moth holes had disappeared altogether.

CHAPTER NINE

Cat Versus Cat

'Ignorant people think it's the noise which fighting cats make that is so aggravating, but it ain't so. It's the sickening grammar they use.'

Mark Twain (*Samuel Langhorne Clemens*), 1835–1910

Some of us, who love cats too much, are severely tempted by the idea that if one cat is enjoyable, two cats would be twice as enjoyable, three even more fun, and so forth. It is as if one cat is not enough. Slowly the numbers climb: one, two, three, just one more, then just one more after that. Cat addiction is not yet labelled a psychiatric illness along with drug addiction, sex addiction and gambling addiction, but perhaps it should be.

Animal hoarding is beginning to be recognized as identifiable human disorder that leads to immense animal suffering. Every now and again cruelty inspectors have to move into a house or flat that is overcrowded with animals, often cats. There are cats everywhere; cat excrement is widespread and usually the cats themselves are suffering from a variety of diseases and disorders.

We cat lovers, of course, usually aren't that bad, but it is possible for us to have too many cats even if we have only three or four felines. How do we know if we have one too many? Signs include if one cat starts to live upstairs, perhaps under the bed, and only comes out at night to use the litter tray downstairs; if one of the cats re-homes himself or huddles under the garden shed all day, however bitterly cold the weather; if there is cat urine, usually sprayed up against a vertical surface; or you come home to find tufts of fur all over the living room.

Cats are only happy if they can have enough personal space. One survey showed that, when indoor cats were within sight of each other (and a lot of the time they were not), they kept a minimum distance of between one to three metres (three to nine feet) apart.

Cats are not sociable like humans

If you think about your life in the past month, you probably spent time at work with other human beings, went out to a bar or coffee shop with a friend, invited somebody to lunch, and enjoyed yourself at the cinema or football stadium with scores of other humans. We humans choose to eat together and socialize together, and our peak moments, whether weddings or degree ceremonies, involve spending time with friends and family.

Cats are not like this. They hunt for food alone and eat it alone (unless they are feeding kittens). They play with bits of string and cat toys but only with humans, not with other cats. You can't imagine a cat being part of a football team but you will often see dogs joining in a game of football with their family. On the whole, cats just don't do that sort of social life.

The colonies of feral cats, where females share nursing duties with their sisters, are usually made up of related cats. These family colonies chase off unrelated cats that venture into their territory (unless they are sexually attractive toms!) and they only live together because that's where the food and shelter is.

Yet we expect our cats to live together at our command. We expect mothers to live with their kittens even though we'd hate to live with our mothers. Or we expect several cats to get on together even though they are not relatives. Without thinking twice, we will add another cat to the household and expect the existing cats to accept the new intruder as a friend.

Sometimes the problem begins after a cat, one of a sibling pair, has died. The owner sees the remaining cat wandering around apparently looking for his dead sister and thinks he must be lonely. He is missing his sister. To the human, the solution is obvious – get him a feline friend. But to the cat this is not so much a solution as a new problem; first the cat loses a beloved sister, and now he is expected to get on with an intruding stranger who is thrust into his hitherto safe home.

CAT FACT: Fights are likely to break out when a new cat is added to a household. One study showed that less than a quarter of the new cats were accepted without hostility. A month after the introduction, about half the cats had settled down to live without aggression but fights were still occurring two to 12 months after the introduction. These later fights were more likely to occur if there had been a fight at the first meeting. This suggests that it is worth introducing new cats gradually and with careful manipulation of scent, including using Feliway®.

Companions or friends?

If you want to know how your cats feel about each other, spend some time looking at the space between them. Close feline friends sleep together, groom each other and sometimes walk together with intertwined tails. They are literally, as well as emotionally, close. Mildly friendly acquaintances may greet each other by nose touching, but they don't sleep together or groom each other. They will probably be able to share their owner's bed or the sofa, but they will keep two or three feet apart from each other. They can pass each other in a doorway, however, without their ears going back.

Indifferent acquaintances keep even more distance. They will choose separate chairs or beds to sleep in. Rather than pass each other in a doorway, one will wait until the other has gone through. Hostile acquaintances can't go through a doorway without showing signs of hatred – hissing, raised paws, ruffled fur, their ears held back. If they find themselves too close, they will often sit low to the ground and glare at each other.

True enemies need even more space from each other. They may live in separate areas of the house, one upstairs, one downstairs, and there will be glaring matches. One cat may ambush and bully the other so that the victim starts living under the bed or takes up life in the garden. Fights also occur. These cats should not be forced to live in the same home.

> ### CAT TIP
>
> If cats cannot get on, one of them should be given the chance of a happier life by rehoming the other. However, if this isn't possible, install a PetPorte cat flap (see p. 114) inside the house. This is linked to a microchip and will only let a particular cat through. Using one of these means you can give a bullied cat a refuge of his own in, say, the linen cupboard.

Sociable and unsociable cats

There is no rule about which cats will like other cats; because cat A likes cat B, it does not follow that he will like cat C. There's been some scientific research into why cats vary so much in their ability to live with other cats. Cats in general are wonderfully flexible – sometimes living in large colonies and at other times leading loner lives on their own – but individual cats in particular may not be flexible at all. Some are loners by nature; it's in their genes.

The other reason for a cat being unsociable is that she wasn't well socialized as a kitten. Just as feral kittens that never meet humans grow up scared of humans, so single kittens that haven't met any other cats than their mother may find it difficult to get on with other cats. Kittens that grow up in a home where they meet not only their siblings and mothers but one or two adult cats as well, have a better chance of adapting to a multi-cat household in later life.

For some cats, the company of other felines is immensely stressful. I used to volunteer in a cat shelter where there were semi-transparent walls between cat pens. Whether a cat was in his bed or out in the run, he could always see the cat next door. Some of the cats didn't mind this, but others were really unhappy about it and I would see them trying to fight off the nearby cat by attacking through the transparent wall. A cat that wanted to be alone had literally nowhere to hide. As a solution to this problem the shelter decided that each cat should have a bed that partly hid its occupant from the sight of other cats.

CAT FACT: Brother and sister cats that grow up together from kittenhood usually stay close all their lives – though there are exceptions, just as there are in human life. It's probably something to do with meeting each other very early on before the age of eight weeks. Scientists who looked at a small survey of cats in a cattery found that related cats slept together and groomed each other, comfortable in each other's company, while the unrelated cats did not.

Cats don't share – they time-share

Given a choice (and this choice is often withheld), cats prefer to use most facilities on their own. Humans, who enjoy eating with others, often post pictures on the web of their cats eating side by side. It looks as if the cats are friends eating together, just as we humans eat together. The pictures, however, may lie. What they show is that, if food is put down in a row of bowls at the same time, cats will eat together rather than go hungry. Some of them may be friends; others are not.

Most cats (and there are always exceptions among ragingly individual felines) prefer to eat on their own. If a row of food bowls were to be kept full and left down all day, it is much more likely that you would see the cats coming one by one to eat throughout the day. They would time their meals so as not to coincide with those of the other cats, preferring to eat at a distance.

The same is true of litter boxes. Obviously cats have to time-share litter trays, using them at different times, but difficulties sometimes arise when a cat decides that he will not use the same litter tray as his companion. It is unclear why this decision is taken. Owners will often say to me: 'But they have shared the same litter tray for years.' My response is simply: 'He has changed his mind about it.' Usually there is no apparent reason for this, but once a cat has decided, his inbuilt persistence means that he is unlikely to change. If you want feline toileting to remain in the litter tray, get another tray and put it in a different location to the first one.

Secrets of feline harmony at home

The secret of a happy home in a multi-cat household is plenty of everything. There should be a litter tray for each cat, and the trays should placed in different locations, not just in a row. If there is any bullying going on, this makes it more difficult for the bully to lie in wait and ambush his victim near the tray area. The most controlling of cats will fail to control access if there is more than one litter-tray location.

Litter-tray hygiene is also important if cats are using the same trays. If any of the cats shows signs of relieving himself outside the litter trays, an extra litter tray should be added. Owners get pretty fed up with having to find space for all these trays and I have to assure them that anything, however irritating, is better than the smell produced by a cat who has started using the back of the sofa as a toilet area. If you have five cats you need five trays. (You might even need six if the cats live indoors all the time and don't get on too well with each other.)

If you are lucky, you clean the litter twice a day and you put down plenty of fresh litter, you may get away with fewer than one tray per cat. However, while you can't have too many litter trays, you can have too few. Location matters, too. It's no good just putting them all in the same place. Trays should be in about three different locations – perhaps the utility room, the upstairs bathroom and perhaps the downstairs toilet, too.

The same arrangements go for feeding locations. Make sure that in a household of more than two cats there are at least two feeding locations (not too near the litter trays). One upstairs and one downstairs are useful. I keep a second feeding location in my bedroom, placing a water bowl and a food bowl on a tray. The tray stops pieces of cat food falling on the bedroom carpet.

There should also be plenty of cat beds, both upstairs and downstairs. Of course, cats will find their own place to sleep but it can do no harm to add a few extra beds. Cardboard boxes, with an entrance cut in one side and with a piece of old blanket inside, will do nicely. Five cats should have a minimum of six different sleeping places (more would be better) so as to reduce the chance of squabbles.

CAT TALE: Just because cats live together without fighting, it does not mean they like each other. Jilly, the illustrator for this book, lives in two old cottages that have been turned into a single home and has four cats. Her cats are not a united family at all. Instead, each individual has a distinctly different relationship with the others. Two, Blaireau and Figgy, are the closest of friends. Figgy (full name, Figglepuss), a three-legged mottled tortoiseshell, is Blaireau's disabled mother. Blaireau, a tabby, cares for her, helps her groom herself, cleans her ears, licks her face and curls up with her. Very occasionally they will even eat from the same bowl.

Even being related does not ensure friendship. The other two cats, Fanny, a grey tortoiseshell and Enki, a sort of fluffy grey, are daughter and mother but do not get on. They may be related but they don't like each other. Besides, Enki does not like any of the other cats. She is a natural loner, a solitary cat who leads a life as separate from the others as she can make it. She sleeps in the top half of one side of the house, guarding the stairs to Jilly's bedroom so that none of the other cats come into her territory.

Blaireau and his mother Figgy sleep near the computer during the day (where Jilly is likely to be) and near the Aga oven at night, the warmest place in the house. Fanny tries to sleep on Jilly's lap whenever possible and at night sleeps in the living room.

Each cat has an individual relationship with the other. Fanny and Blaireau, for instance, quite like each other but Blaireau hates Enki. All four – Figgy, Blaireau, Enki and Fanny – live without conflict in the same house by carefully not intruding on each other's preferred space.

How many cats are too many?

How many cats should you have? As cats are so individual, it depends on the individual cats. Sometimes more than one cat is one cat too many. There are some cats that really need to live alone to be happy. For them, even one extra cat will be stressful. Other cats are natural bullies and can really upset, or even harm, their feline companions. However, two sibling cats, brought up together from kittenhood arc usually compatible.

Some experts recommend keeping no more than two cats unless the house has more than three bedrooms. It is not just the floor space that counts, it is the design of the house. An old rambling house with back and front stairs, or a house with two separate living areas, can have more cats than a very open-plan modern house. De-cluttered houses with nowhere to hide are bad for feline harmony.

If you work on the assumption that you need one litter tray per cat and maybe one more, all in different locations, and that more than two cats will need two feeding locations, you can see that you will need a largish house. Then consider the cost of booking them all into a cattery, the vets' fees, and the time spent grooming and caring for them properly, and there is another reason for not filling the house with cats. Cat collectors usually don't do cats any favours.

The more cats there are, the more likely it is that you will have trouble between them. A well-known cat behaviour counsellor did a survey some years ago that showed that as the number of cats living together rose, so did the likelihood of spraying in the house. Indeed, eight out of ten households with seven cats or more had experienced the pungent smell of a cat marking its territory indoors.

CAT FACT: Aggression between cats was most likely in homes with three or more cats, according to the cases referred to the UK Association of Pet Behaviour Counsellors.

When household cats fall out

Any group of cats inside a house is a network of indifference, friendship and dislike. Usually once the cats have settled down together, relationships remain stable. Cats that have lived together for years often manage to work out a way of tolerating each other – or leave home! – but sometimes a change upsets the equilibrium. The introduction of a new cat, the disappearance of an existing house cat, the arrival of kittens, or a change in routine can adversely affect household harmony.

Even a visit to the vet may ruin a hitherto friendly relationship between cats. Smell is the key to this. Cats identify friends and foes by their scent. An important part of the household harmony is the family smell – a smell made up of a mixture of humans and all the cats. If that smell is disrupted, all hell may break loose.

Cats that visit the vet are usually placed on an examination table and handled by the vet. The examination table will have been disinfected after the previous patient and the vet's hands should smell of disinfectant and soap, if proper hygiene is in operation. All this adds up to a distinct veterinary smell that most cats loathe. They know only too well that the vet is likely to jab them with needles, force open their mouths to examine their teeth and even assault their privacy by sticking a thermometer up their backside. A cat that comes back from a visit to the vet not only smells strange rather than familiar, he smells like the enemy! Thus he is often attacked by a previously friendly companion cat.

Similar upsets can occur when cats go missing for several days and return smelling of a different household. Or hostilities may break out over medication or something like shampoo that is applied to one of, rather than all, the cats. The trick is to make sure all the cats smell the same by wiping a tiny dab of the medication or shampoo on all the cats. To stop potentially harmful substances being licked off, dab it at the top of the cat's head or the nape of his neck where he cannot reach it.

CAT TIP

Get all your cats vaccinated at the same time. This means that you can take all of the cats on the same visit to the vet's surgery. They will, therefore, all smell of the vet on their return. If only one cat requires veterinary treatment, try to bring him home and put him in a room on his own for a little while for the vet smell to vanish. It may help to rub some of your own favourite perfume on your hands and then stroke him – so that he smells of your perfume, a smell familiar to the other cats, rather than smelling of the vet. If the feline relationships within your home are fragile or verging on hostility, just take all the well cats with the sick cat. Ask the vet to handle them all before you leave.

Cats cannot forgive and forget

Cat squabbles easily escalate into full war if you don't make an effort to allow the cats to lead separate lives. Cats can't say sorry. Dogs can. Dogs have a range of what are called appeasement behaviours – lifting an apologetic paw, lowering their body and crawling, lying on their backs or turning their gaze away. When they do these appeasing moves, the other dog will accept their apology and leave them alone.

Cats don't do apologies like that; they literally don't have the body language of apology. If they fight, there is no way they can make up afterwards. So if your cats are squabbling, you must take action as soon as possible to prevent it getting worse.

The best way for their tempers to cool off is to keep them away from each other in separate rooms for a few days. Then gradually re-introduce them, making sure there are separate eating places, separate locations for litter use and separate beds. It is vital to have plenty of feeding places, litter trays and litter-tray locations and beds. Cardboard boxes with entrances and a window will supply a place for a timid cat to hide. A Feliway® diffuser plugged into the room where the cats spend most time also helps restore harmony. It exudes a calming scent, a little like the scent of friendship.

Sometimes this just doesn't work and squabbles become real fights. Stress can create conditions in which one of the cats may fall ill with cystitis. If two cats really can't get on with each other, rehoming one of them may be the only fair thing to do.

Usually there is a bully and a victim. No victim cat should have to live in constant fear of attack.

CAT TALE: Tanya and Amber, two spayed sisters, lived together for eleven years without any difficulty. Beautiful pedigree Turkish van cats, they were never close friends but they were tolerant and mildly affectionate towards each other. Suddenly, an enormous fight broke out. The trigger for this was probably what experts call transferred or redirected aggression. A strange cat intruded into their garden when both cats were sitting peacefully in it. Fired up by this, Amber turned her aggression not onto the intruding cat that had fled, but onto her sister, Tanya. Tanya wet herself with fear. Growling, howling, hissing and glaring continued whenever the cats saw each other.

A visit to the vet confirmed that no wounds had been inflicted. However, the relationship had completely broken down. Tanya, originally the victim, spat and growled at Amber warning her to keep her distance. Amber growled and spat back, sometimes backing Tanya into a corner.

I suggested a three-stage treatment. Stage one consisted of complete physical separation between the two cats for a cooling off period. When the oweners, Pam and John, were out of the house, the cats were put in separate rooms. When they were in the house, some visual contact was allowed, using a crate big enough to contain bedding, litter tray and water. The cats took it in turns to be in the crate or pen. During this period, a Feliway® diffuser was installed in the living room to exude soothing smells. This was topped up by a daily Feliway® spray.

Stage two of the treatment was to let the cats be together without the pen, while there was a human there to supervise. Pam and John were devoted cat owners who could read their cats' moods. They moved to stage two when they felt the cats were able to be in the same room without fear or aggression. They even cancelled their holiday to make sure that they could be at home to supervise the cats.

Stage three was to go back to the original situation with both cats living in the same house with free access to all areas. After a few weeks, John and Pam allowed this. A Feliway® diffuser was renewed monthly in the living room to promote calm and only dispensed with when Pam and John were confident the cats could be together without aggression.

Introducing a new cat

Adding a new cat, especially an adult cat, to the home can disrupt the feline set-up. From the point of view of the existing cats, this is an intruder into their territory. They don't want a stranger in their home. The newcomer smells wrong and they just don't like this feline at all. If the cat is just put down among them, they may well attack. Worse still, they may start spraying inside the house as a way of marking their antagonism towards the newcomer.

So introductions should be taken slowly. A Feliway® diffuser can be installed a week before the introduction so that its soothing scents are already in the air. It is a good idea to install the new cat in a spare bedroom for a day or two so his scent can begin to waft out into the household.

Bedding should be swapped between the existing cats and the new arrival, so that the new cat starts acquiring the family scent. Pet the old cats, then the new one, so that your hands also take the scent from one to the other.

When adding a new cat to an existing single cat, it is probably best to add one of the opposite sex. If your existing cat is male, add a female, or vice versa. Making sure that the cats are roughly the same size will also help; a very large cat, such as a Maine Coon, is more likely to injure or frighten a small cat if a squabble breaks out.

Do everything you can to allow both cats to avoid each other. Make sure they have separate access to food and litter. Remember, keeping a distance is the way that cats maintain harmony. If they are forced into each others' company, they may have to fight. Your aim must be to stop a fight happening in the first place, as once they have fought they may not ever be able to forgive and forget.

Kittens may be accepted more easily, but even so this acceptance is likely to be grudging to begin with. Existing cats rarely mother kittens; mostly they show their disgust by ignoring them, or hissing at them if they get too close, or refusing to acknowledge their existence. It is a good idea to introduce a kitten in a crate that is big enough for a bed and a litter tray. There he will be safe if you leave him with the other cats. The other difficulty about introducing a kitten is that you cannot be sure of his personality, whereas a good rescue shelter should know which of their adult cats can tolerate other cats.

CAT FIGHTS

If there is fighting between cats, the following questions will allow you to assess how bad it has become.

1. Does the victim need veterinary treatment?
2. Are there wounds – bitten ears, broken skin, blood, etc?
3. Is one cat spending less time in the house because he is afraid to come indoors?

4. Is the victim spending his time hiding somewhere, like under the bed?
5. Is the victim no longer grooming himself?
6. Is the victim too frightened to eat when the other cat is in the same area?

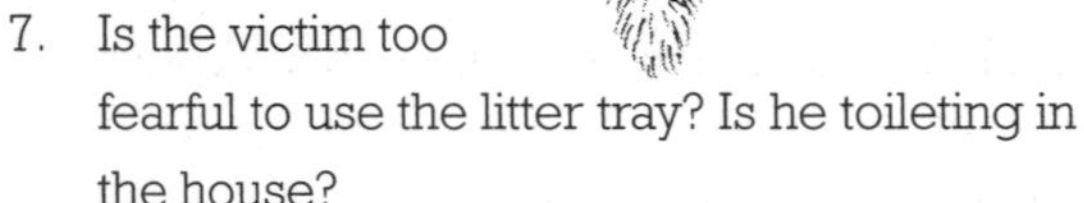

7. Is the victim too fearful to use the litter tray? Is he toileting in the house?
8. Is the victim no longer willing to approach you because of fear of being attacked by the other cat?
9. Do you come back into the house to discover signs of conflict?

10. Is the aggressor patrolling an area of the house so that the victim only has a small area of living space?
11. Is one of the cats spraying in the house to make territorial marks?

If you answer yes to any of these questions, this is serious aggression. If you answer yes to more than three questions, you must rehome one of the cats or get the help of a cat-behaviour counsellor. The victim must not be allowed to continue to suffer pain and fear.

Introducing a new dog

Humans can be even more selfish than cats; they will introduce a dog into the household without ever considering the welfare of the existing cats. Yet a single snap from a large dog is death to a cat. Even a small dog can maul or completely traumatize a cat. A cat's safety and happiness should always be paramount over the human desire for a dog. Cats that have lived with dogs before should be able to cope, but cats that have never lived with a dog may be extremely frightened.

If you must add a dog to your feline household, choose the breed carefully. Terriers, even small ones like Jack Russells, have strong predatory or hunting instincts. So do greyhounds, lurchers, and other chasing breeds. It's not their fault; they are bred to chase after small, furry moving targets. Greyhounds can and do live with cats, but I, for one, would not risk it. Very careful introductions are essential if you must have a greyhound.

There is one seriously important thing to know about introducing dogs to cats. Never let the dog chase the cat. Ever. Once a dog has chased a cat it sees it as prey, not a friend. And a cat that has been chased will probably never feel safe again around that particular dog. It will be difficult, even impossible, for a friendly relationship to develop.

If you have cat and want to add a dog, choose a puppy from a gentle gundog breed, such as a Labrador or a spaniel. Introduce it to your household cats as a puppy in a crate. When the puppy is not in its crate it should always be on a lead. The aim is never to give the puppy an opportunity to chase the cats. Bearing in mind the importance of breed, an adult dog from a rescue centre is a possibility – if it has grown up with cats. Good rescue centres will have checked up on this. If the rescue centre doesn't know for sure, find another animal shelter that does. Take no risks with your cat's safety. For the first couple of weeks the dog must be on a lead always, just to make sure a chase never happens.

Scent is the other precaution you can take. Swap bedding between cat and puppy or dog in advance. A good animal shelter should allow you to add a piece of cloth to the dog's bed (to take home to your cat's bed) before the dog is taken home. Install a Feliway® diffuser a week before the dog's arrival.

The cat will also need places out of reach of the new dog where they can feel safe to sleep, eat and toilet. This means placing beds, food bowls and litter trays at table level rather than floor level. The cat must always be able to avoid the dog. Installing a cardboard hiding place will help. A stair gate, through which the cat can pass when the dog cannot, ensures that the cat always has a place where he can retreat.

Life stinks when things go badly wrong

When the telltale smell of cat pee starts wafting through the house, do something. Sometimes you will be lucky and just improving the litter tray numbers, types and locations will help. At other times it is a sign that this is an unhappy home.

Possibly one cat is so terrified of being ambushed on the way to the litter tray that he has started peeing in the corner of the bedroom or behind the chest of drawers. It will be a secluded place. Your cat is becoming frightened of his companions. He may be being attacked as he uses the litter tray, or one of the other animals has started ambushing him as he makes his way to the litter tray in the utility room. So the cat now feels safer doing it upstairs out of the other animals' way.

The other alternative is worse. In this case the smell comes from one of your cats, who is marking his (or even her) territory. In this case, the urine marks will probably be up against a vertical rather than a horizontal surface in a noticeable, rather than a hidden, area. The spraying cat feels his territory is being invaded. He is very anxious and his way of coping is to leave smelly messages with urine.

We can't be sure, but the messages probably read: 'Keep out', or 'Trespassers will be attacked', or even 'Angry cat passed here'. When this happens, don't hang about. Do something. How to deal with this situation is addressed in the next chapter.

CAT TALE: Dougal, a cat nicknamed Puddles by his fosterer, was a stray cat rescued from the streets by a charity that neuters and replaces feral cats. He was black and white, probably about three years old, and absolutely terrified when he came to live with his fosterer. He began peeing on the carpet a few days after his arrival in the foster home. He used the litter tray for solids, however, showing that he knew what it was for. I suggested to his foster mother Sheila that she put out a second tray in case he was one of those cats that didn't like to pee and poo in the same box. It worked for a few days but then he started peeing on the carpet in another room. Sheila moved the litter tray to that location and he started using it reliably. Obviously Puddles not only wanted separate litter trays, but also two separate locations. After a week or so he was rehabilitated to life with two litter trays and found a home of his own.

CHAPTER TEN

The Smell of a Cat Crisis

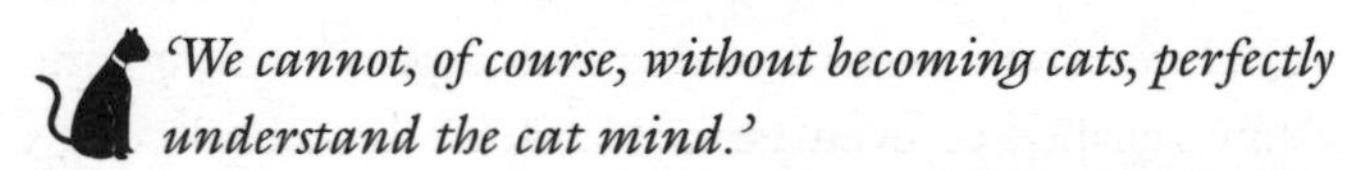

'We cannot, of course, without becoming cats, perfectly understand the cat mind.'

St George Mivart, 1827–1900

If your cat starts peeing in the house, you need to respond fast. Very fast. There is nothing more disgusting than the smell of cat urine ingrained in the furniture and floor of a house. Once it is well established it is almost impossible to exterminate, except by full-scale house redecoration.

The first stop on the road to a clean house is to take your cat to the vet to make sure that she is not suffering from an illness such as cystitis. Kidney disorders and other illnesses can mean that a cat stops using her litter tray. Cats with arthritis or a back injury that have to use the stairs or go across a polished wood floor may not be able to reach the litter tray in time. Cats that are very old or very ill may be suffering from elderly incontinence.

A particularly common cause of urine in the house is cystitis or feline lower urinary tract disease, known as FLUTD. A cat with cystitis often can't get to the litter tray in time and has to pee in the house. If you have a litter tray in the house, you will also see repeated small deposits of urine – about the size of a ten-pence coin – instead of normal large deposits. Cats normally urinate two to four times a day and defecate once or twice daily. Exaggerated use of the litter tray, like failure to use the tray, is a sign that something is wrong. Stress plays a part in FLUTD, and so territory troubles are relevant for this illness.

As a cat behaviourist, I have discovered that human beings have fixed expectations of their cat. They expect her to use the litter tray reliably, or even go out and use somebody else's garden rather than their own. They expect the cat to put up with dirty litter or even a tray with very little litter in it at all. When the cat behaves otherwise, they get immensely upset and *are unwilling to make new arrangements.* I have to remind them that there is probably nothing worse than a house smelling of cat pee. Assuming that a health check with the vet shows no signs of illness, attention must next turn to the litter tray.

CAT FACT: Between 7 and 11 percent of cats that are handed into rescue shelters in the UK are given up because of behaviour problems, and about one in a hundred cats are put down by vets because of behaviour issues. The most common problem referred to behaviour counsellors is peeing or pooing in the house, followed by aggression to people, then aggression to other cats.

Well-maintained litter trays, like bathrooms, are essential for happiness

We humans know how stressful it is when we need a lavatory and can't find one. It is just as stressful for cats. Indeed, worse. They can't ask for what they need. Imagine what it is like to have to use a toilet that smells of the previous user, or even smells of the previous two users. Think how stressful it must be for a nervous cat to have to worry about being ambushed on her way to the litter tray; or, when she gets there, to find that the familiar litter has been altered to something she doesn't like at all. Worse still for the cat, to get there and to find it stinks (in its delicate nose) of heavily smelly deodorant.

You, the human owner, want just one litter tray, with inexpensive litter in it, placed where you like it in the corridor or in the conservatory surrounded by glass. You only want to clean it once a day even though there are three cats using it. Is this fair?

What your cats want are litter trays that are in secluded places where nobody can look in through the windows to see them on the toilet. They do not want a litter tray in a place past which the whole household tramps. They don't want inexpensive litter in huge pieces, that will be like shifting cobble stones when they dig. Most cats want small-grained clumping litter.

Have you ever noticed that your cat insists on using the litter tray just when you have cleaned it? Did you sigh with exasperation? If so, you missed the point. Kitty was trying to tell you something. She likes *clean* litter, not litter full of pee and poo clumps.

Cats have highly individual preferences for litter trays. They get used to a routine with a type of litter, a type of litter tray, and a settled location. Change the litter, or litter tray, or its location, and you may find your cat protests by refusing to go inside it. The result is pee or poo in the house. Go back to the original arrangements.

Cats with cystitis sometimes experience a burning sensation when passing urine or an ill cat may be in pain while on the litter tray. Cats can think this has something to do with the litter tray and decide to avoid it, so it's always worth experiment by installing a new tray. If the cat uses this with relief, you have solved the problem.

Sometimes, just to keep you on your toes, a cat that has happily used a litter tray for both pee and poo for years, decides it will no longer do so. You cannot change this. Put down a second litter tray and be grateful for finding an easy solution to poo in the home.

If your cat does not have a litter tray at all and has started using the house instead, install one immediately. She may be being bullied by neighbouring cats, and journeying through enemy territory two or three times a day to get to her toilet has perhaps become too much for her.

CAT TALE: Jonesy, a neutered tomcat from a rescue shelter, would never go outside the house but reliably used the litter tray indoors. Libby, his owner, decided that the litter tray was looking shoddy and bought a bigger, better one. She put it in the same location and took away the old tray. The same night Jonesy relieved himself on the living-room sofa. The next day he did the same thing. I advised putting back the old tray. Luckily, it was still in the dustbin so she pulled it out and reinstalled it. Jonesy used it immediately. Problem solved.

Cat urine in your home – the three-fold solution

There are three essential steps for persuading a cat to stop urinating in the house:

- Clean up properly.
- Use Feliway® or your cat's scent to make the areas smell right. Put food near the urination spots to deter future use.
- Identify and deal with the causes; whether these are litter-tray problems or stress that have led to territory marking by spraying (pee) or middening (poo).

These three steps are often misunderstood. All cat owners follow step 1 and try to clean up but, alas, they often do so in a way that encourages rather than discourages the problem (see pp. 209–12). Most good vets will suggest step 2, using Feliway®, but few give proper instructions on how to do it (see pp. 212–13). Only a very few cat owners and vets move on to step 3 and really deal with the causes of spraying. If you don't apply your mind to why your cat needed to spray, you have little hope of permanently stopping the problem.

CAT FACT: Pheromonotherapy is a major breakthrough in the treatment of cats that spray in the house. French researchers isolated the scent that cats leave when they rub against their friends or items in the house. It is a scent that promotes a relaxed mood in a cat. You could say that it smells of home and loved ones. Cats don't spray where they rub, so this artificial scent, Feliway®, discourages them from spraying.

Location – where is the cat going?

Some humans feel personally insulted if they come home to find a mess of cat pee or poo on the bed. The smell disgusts them. They feel that the cat is treating them like shit. They take it as a personal insult, a message of ill will from their cat.

Well, it is a message, but it's definitely not an insulting one. Many people believe that pee in the house is a cry for help. Others believe that territory marking with pee and poo is one of the natural ways that cats organize their personal territory. (Other territorial marks include scratching and facial rubbing.) It is therefore more likely they will react with pee or poo if their territory is changed in any way. Anything new, uncertain, or threatening will prompt them to spray.

Sometimes when a cat pees or poos in the house it is just a litter-tray problem. How can you tell? Well, the pee or poo may be just outside the litter tray, or it may be in a secluded place like the corner of the utility room, upstairs in the bedroom corner (to where the cat has retreated) or behind the living-room sofa.

The cat no longer likes the litter-tray arrangements and is finding somewhere else to go. The pee will be on a horizontal not a vertical surface and, as I have said, in a secluded part of the house. The cat is not marking its territory with messages; she is just finding somewhere else to use as a toilet. Dealing with the litter-tray problems (as perceived by the cat) should put an end to the urine deposits.

CAT FACT: Research into scent marking with poo has come up with conflicting results, but it is possible that wild cats sometimes leave uncovered poo as a way of territory marking. Pooing in the house, therefore, may be just a question improving the litter-tray arrangements, or it may be rather similar to urine spraying – a sign that the cat is feeling stressed and is marking her territory.

Reading the clues of pee and poo

However, cats that are perfectly happy with their litter tray may still pee in the house. If your cat is backing up against a vertical surface and letting go a stream of urine from under her tail, this is spray-marking. It is leaving a scent message. You will see that the urine is almost always on a vertical surface. Entire tomcats spray their territories frequently; for them, it's a male thing and the only way to stop them spraying is to give them the snip.

However, neutered toms and female cats also spray. One estimate is that one in ten of neutered males and one in twenty neutered females will spray. It's natural feline behaviour. When they do this outside in the garden it's no problem for humans, but they may start spraying inside the house if they feel their security or their territory is under threat. Occasionally they will territory mark by leaving poo instead of pee.

Location is a clue to what is going on. Territorial urine marks will often be at the edges of the home – below a window, at a doorway, or on an indoor windowsill. When the cat looks out of the window or sniffs at the doorway, she may see or smell another cat in the garden. She may even smell another cat's urine, where a neighbouring cat has sprayed on the doorstep.

Once cats start spraying in the house, they will continue. They regularly top up their spray marks. They often then start spraying in more locations. So what starts with a spray mark near the French windows as a reaction to seeing a new cat in the garden, continues with the cat marking beneath all the downstairs windows, marking the door of the cat flap through which the new cat can be seen or smelled, and then going on to spray the inside of the front door. If it isn't treated, spraying gets worse.

Furthermore, the cat may start spraying on other objects in the home. Cats that have previously lived near a mirror or reflective glass without any trouble will, if they have become anxious about territory, start spraying near the glass. When they see their reflection, they think the image is a strange cat in their home. Home items, such as radiators or even electric kettles, that give off a plastic or a metal smell when they heat up, get a jet of urine too.

Some cats will even spray on the shopping. This seems crazy until you consider how most of us shop. We walk to the door with our shopping bags, put them down, and then put the key in the lock and open the door. The bags have been put on the doorstep, and if a neighbouring cat has sprayed there, the bags will smell of *his* spray. So when you bring this smell into the home, your cat feels compelled to add her spray mark to cover up the hostile scent.

To make things even more confusing, the litter tray may start feeling unsafe for them too. Often the litter tray is kept in the utility room near the cat flap. If a neighbouring cat is intruding into the house through the cat flap, it goes past the litter tray. The frightened house cat starts avoiding the cat flap and therefore also avoids the litter tray.

So now you are left with spray marks *and* ordinary pee marks. At this point you may need the help of a cat behaviourist who can work out what is going wrong. In the meantime, you need to clean up – thoroughly.

CAT FACT: In order to understand why and when cats scent mark, scientists have studied the life of farmyard feral cats. In one study of a farm cat colony living in a barn, the resident tomcat sprayed throughout his territory. About one sixth of the spray marks happened after he had had an aggressive encounter with a strange cat. This fits in with the idea that household cats start spraying when they have hostile encounters with another cat.

Cleaning up cat pee and poo

Humans usually clean up cat pee the wrong way: they will use a cleaning fluid that has bleach or disinfectant, often with the smell of pine or lemon. Once they have rubbed the area with this, they can no longer smell the cat pee. Instead they smell the comforting (to us) aroma of clean pine or clean lemon.

Wrong. Your cat can still smell the pee below the disinfectant fragrance, and, what is even worse, the bleach or disinfectant smells to them like cat pee anyway. So for your cat there is the smell of his original spray mark and then on top of it is the smell of a foreign intruder (actually the bleach or disinfectant). She feels compelled to spray a new 'Keep out' message on top of this intruder's scent. The cat is now even more anxious to make sure her spray marks are up to date and so she sprays repeatedly.

Cleaning up cat pee needs biological washing powder or liquid in a warm one-to-10 solution of water. Scrub, then rinse with clean water and leave to dry, or use a hair dryer to dry it. Then, using a plant sprayer, spray surgical spirit on the site. Rub this in with a nailbrush or wipe with a cloth.

Curtains and furniture covers need dry cleaning. Bed linen must be washed. Carpet that has been repeatedly sprayed on may need renewing. Radiators, after cleaning, may be repainted with metal paint.

CAT TIP

If your cat is spraying in the house, turn off all scented plug-ins. A cat likes her home to smell of herself, her familiar humans and her familiar animal companions – a kind of friendly group scent. Cats' noses are extremely sensitive and commercial scent devices are strong. Wafts of scent may smell like roses or lily-of-the-valley to us humans, but to cats they may give the impression of foreigners in the home.

CAT TALE: Perkins and Warbeck were two pedigree Siamese cats. Their owners, Anne and John, knew nothing of cats but decided cats would make easy pets in their retirement. They would have liked a dog except that they planned to spend a lot of time on the golf course. A breeder told them she had a nice pair of young cats that were three months old.

The first six months went well with the cats living in the house and using a litter tray in the utility room near a cat flap. Then Anne and John began to find urine marks in the house. They scrubbed these conscientiously with disinfectant and installed several scented plug-ins in the house. The cats continued to offend, so they were moved out of the main house to the adjoining conservatory. For the next year John and Anne cleaned up the conservatory almost every day, and kept scented plug-ins at full go. When I walked in there I could smell nothing but the pungent fragrance of artificial honeysuckle and disinfectants. I have never been in a house kept so clean and free of urine scent – free of scent to human noses, that is.

By the time I came on the scene, John and Anne were really fed up. The 'easy pets' had turned out to be a nightmare. For two years they had struggled with the problem. Unfortunately, almost everything they had done, they had done wrong. The cleaning regime and the smell of disinfectant simply encouraged the cats to top up their spray marks. The local cats paraded up and down past the conservatory glass, looking in and glaring at poor Perkins and Warbeck. There was literally nowhere for them to hide from the glares. The plug-ins probably stressed them out further.

I drew up two possible plans for the cats. One was to clean up properly, use Feliway® spray lavishly and a Feliway® diffuser instead of scented plug-ins. It would be necessary to hide the sight of the neighbouring cats either by taping cardboard to the conservatory glass, or frosting the glass. Alternatively, the cats could be moved back inside the main house, and attention paid to blocking off the sight from the windows.

The other alternative was to just rehome them. John and Anne had never been great cat lovers, just people who thought cats would be easy pets. For them the cats had been nothing but trouble. Perkins and Warbeck were beautiful cats who would easily find another home and be happier in a country area without too many cats nearby, where they could get on with enjoying the outside life. John and Anne decided to rehome the cats. I felt it was the right choice for the happiness of both humans and cats.

Using pheromone spray the right way

A synthetic pheromone spray is available (see p. 244 for full details) that smells of a cat's facial glands – the area that deposits scents by rubbing. Where cats rub they do not spray, so by using this you are making the spray site smell like a face-rub site. Used correctly it will prevent or reduce a cat spraying. However, many people do not know how to use it properly.

After cleaning and using surgical spirit to clean a marked site, leave the area for 24 hours before you use the spray, otherwise the surgical spirit interferes with its efficacy. The best way to do this is to shut the cat away from the cleaned site. If you can't shut her away, then temporarily cover the site for 24 hours with cling film and use the spray on top of the film. Take away the cling film after 24 hours and then spray on every spray mark site once or twice a day for a whole month. Be lavish, not mean.

If you can't afford this you can use the cat's own scent. Take a soft piece of cloth, like a handkerchief, and wipe it round your cat's cheeks and chin. (If you have a really scared cat, this may frighten her so much that you will *have* to use the spray.) Then wipe this cloth on the cleaned spray sites, putting the rub scent in the place of the pee scent.

You can also buy pheromone plug-in diffusers, but in this situation these are not as effective as spraying directly onto the marked site. Use diffusers as a back-up: they are particularly useful for exuding a calming scent if you have moved house with your cat, during any household disturbances (such as building work), when introducing a new cat or dog to the household, or helping prevent aggression between two household cats that have fallen out.

As well as spraying with this daily, put more protection in place by putting cat food near the spray site. You don't have to have bowls of food all round the house, instead, take some stiff cardboard and glue half a dozen pieces of dry food to it. Leave rectangles of this where the cat has sprayed. Cats do not spray where they eat and the odour of food will deter them.

This method works not just for sprayed urine but also for territory marking with poo.

Preventing territorial troubles from outside cats

You have now cleaned up; you have used the pheromone spray to make the spray site smell like a face-rubbing site; now for the third step: dealing with the cause of the trouble. Something stressed out your cat and prompted it to mark its territory. Deal with it.

Territorial anxiety occurs when cats have to share their hunting range, the area outside the home, with too many other cats. For your cat, too many cats may just mean one, single, hated local cat. Remember that cats don't see walls or fences as boundary markers, their own hunting ranges are a series of paths that go over or under these human barriers. Your cat may have to pass through several gardens to get to a suitable latrine or to reach the field with the best mousing ground.

Cats can usually share territories without too much trouble by using time-sharing or by keeping out of the way of other cats to avoid conflict. If they feel threatened by a cat intruding too far into their area, though, they will either flee or fight. After such an encounter they may also feel the need to pause and spray-mark their territory.

CAT TIP

Invest in a long-distance water pistol and try to set up an ambush of your own, possibly from a bedroom window. If cats realize the water bursts come from you personally, they will only avoid the garden when they see you in it. If you can ambush them without them seeing you, they may conclude that entering an empty garden is also dangerous.

Country cats probably have enough space for these encounters to be rare, so do cats in very built-up towns where cat keeping is less common, but in surburban areas or housing estates there may be scores of households with cats. Some of the neighbourhood cats will be more territorial than others, and there may even be a local feline bully. If you live in an area with a bullying cat nearby, you may find either that your cat comes home with missing bits of ear or she simply stops going out at all. Then she starts spraying 'Keep out' signs inside the house.

You can make your cat feel more secure by making sure she has a litter tray inside the house. A latrine in your own garden will also mean your cat does not have to pass through enemy territory to find a suitable place to go to the toilet. Some people also believe that cat latrines in the garden, which smell of the resident cat, will keep out strange cats. So if you have a hedge, or shrubs, that are at your garden border, make latrines under them (see p. 117 for instructions).

Next, look carefully at your garden and the front and back of your house. Are there steps, doorways, or plant pots against which neighbouring cats may be spraying? For instance, if intruding cats are spraying on the front door's outside steps, the aroma will waft into your house and your own cat may respond by spraying the inside of the front door. Clean the doorsteps if you think a strange cat may have sprayed there or a dog has lifted its leg there. Move any plant pots further away from the door so that any intruder cat spray cannot be smelled from inside the house.

There is also a common garden plant, the box tree or *Buxus sempiverens,* that in warm weather naturally smells of cat pee. This in turn will attract other cats, including your own, to spray on it. If you have one planted just near your front or back door, consider replanting it further down the garden.

CAT TIP

To help your cat feel secure in the garden, make sitting places for her where she can glare out and intimidate neighbouring cats that are thinking of intruding into her space. Your cat's back should be safely to the house, so that other cats cannot get behind her. A slightly higher sitting place, possibly in a tree, will allow her to look down on any neighbouring cats and trade glare for glare from a superior position.

Does your cat feel safe in her home?

Are the neighbouring cats sitting on the windowsill and peering in, or sitting outside the conservatory glaring through the glass? They may even be staring at your cat from the top of a neighbour's shed, or sitting on your garden wall the better to threaten your cat from above. Sometimes, in areas with a huge number of cats, there is almost a line of neighbouring cats parading up and down the fences and walls, peering in the windows from flat roofs, and sneering at your cat through the French or conservatory windows.

Glaring is one way that cats threaten and bully other cats. You need to think of a method by which you can stop them sitting there – using scrunched up wire netting on a wall or roof or planting prickly shrubs might do the trick. If this isn't possible, think about fencing that will keep out all intruders. A fence needs to be 1.8 m (6 feet) high with a horizontal overhead section pointing outwards (to keep out intruders) or inwards (to keep in your own cat). It may be necessary to have both an outward and an inward overhead section (see p. 125). Talk to your neighbour and local planning office about this.

Opaque fencing, such as wood or plastic, will keep other cats out of sight. It may be possible just to add something to an existing fence or wall, if the problem is not intruders but merely cats sitting nearby and eyeballing. A fairly flimsy bamboo fencing panel, attached to a more conventional wooden fence, may do the trick. Or plant a large, fast-growing conifer.

Deterring cats from intruding into the house

However relaxed a cat, she will probably object strongly if strange cats start coming through the cat flap into her home.

Home is the core territory where a cat *must* feel safe at all times. If she doesn't, she will either take up life upstairs or – worse from your point of view – start spray-marking inside the home. The telltale smell of cat urine will indicate how stressed she is feeling.

Sometimes a few sensible precautions will stop the intruder. A long-distance water pistol can encourage a visiting pet cat not to intrude into the garden. But stray or homeless cats sometimes live by raiding other cat's food or even using their home as shelter. Remove the food bowl from the downstairs kitchen. If food is not available, hungry strays are less likely to intrude. If you wish to leave food down for your own cat during the day, leave it upstairs or in a room furthest from the cat flap.

Stray cats, however, may also be in need of warmth, shelter and human care. These cats are truly pathetic. Local cat charities may be able to help you trap a stray and help him find a new home. You will also be removing a source of trouble and fear from your own cats.

However, some visiting cats are neither hungry nor homeless. To deter the well-fed intruder, who has decided your home offers a better class of cat food or warmer central heating, try to set up an ambush. A small jet of water, just as she enters through the cat flap, may be enough to stop her visiting again. Or visit her owner and, with care and tact, try to work out a way of time-sharing so that you let your cat out only when the intruder stays at home. Cat lovers in the same street are often quite friendly people.

Don't leave the cat flap open all the time. Cat flaps should always be shut at night, to keep your cat safe from traffic, and this also gives her eight or nine hours safe from neighbourhood intruder cats. If all else fails, shut the cat flap 24/7 and only let your cat out when you can bodyguard her.

CAT TALE: Roofer is a beautiful neutered tomcat with grey-brown plush hair and white markings. His character is feisty – a polite way of saying that strangers mustn't intrude on his space. Luckily Sue, his owner, is a perceptive woman who is interested in animal behaviour. Roofer lives with Poppy, a long-haired cat rescued from the streets. Roofer is not fond of other cats. When Poppy was first introduced to him, he had initial difficulties in accepting her, but now they can share the house without any issues.

Roofer's troubles, or, to be more accurate, Sue's troubles, started when they moved home. A black cat from a nearby multi-cat household started intruding into Roofer's garden and even through the cat flap into the kitchen. Roofer's reaction was to spray-mark his territory, putting up 'Keep off my land' signs. Sue sent me the house plan, and it was clear that Roofer had sprayed wherever he thought the black cat might intrude.

He sprayed in the corner of the living room just under the window, from where he could see the black cat coming up the drive. He sprayed in the conservatory where the black cat would look in through the glass and he sprayed near the cat flap. He also spent a lot of time patrolling his territory, watching obsessively from the front window, or going out into the garden to see the black cat from a perch on the garden fence.

There were several fights, after one of which the vet suggested calling me in. I suggested a proper cleaning regime, lavish use of Feliway®, and blocking off sight of the other cat wherever possible. Sue, who understood exactly what was needed, also used cloths which had Roofer's scent on them, placing these on some of the old spray sites.

Better fencing helped keep out the black cat some of the time, but then Roofer had a bit of luck. The black cat, who was not happy in his multi-cat household and was roaming around looking for other places to live and eat, was adopted by a cat-loving neighbour. Happy at home, he stopped his intrusions into Roofer's territory.

Other outdoor territorial threats

Other garden intruders that can terrify your cat are neighbouring dogs (if you live in an open-plan estate), foxes, badgers and other wildlife. Foxes will attack small, young, elderly or already-wounded cats but are less likely to take on a big healthy tomcat. However, they are threatening to all cats so it is a mistake to feed them in your garden if you have a cat. Just their presence in the garden is enough to worry a cat and to prompt her to spray the edges of her territory, such as doors or windows from which the frightening wildlife can be seen.

It's worth getting some advice from wildlife specialists on how to make your garden less attractive to wildlife. Sometimes installing a security light, which will come on at night if any wild animal passes through the garden, will deter foxes. Make sure that dustbins are covered so that there is no food waste to attract them. Keeping your cat safe indoors, once the light starts failing, will also help to make moggy feel more secure.

Windows and glass doors can be blocked so that your cat can't see the worrying wildlife (or glaring neighbouring cats) outside. This is important if your cat is spraying close to the windows. Fix pieces of cardboard with tape to the glass and, if you think this is helping, consider treating the glass with an opaque spray, of the kind used on bathroom windows, for privacy. Don't forget to cover up the cat door if it is made of transparent plastic.

Human intruders and threats

Relaxed cats never worry about visitors in the home. Nervous cats, however, can be extremely anxious when guests come to stay, or when there are parties in the house, or even just when people turn up for a meal. These cats need a place of retreat – normally an upstairs bedroom. Make sure your cat has food, water, and a litter tray if the guests are going to be there for several hours. It is also unfair to expect cats that are unfamiliar with dogs to cope with visiting dogs. A nervous cat should be put in a cattery for the duration of the dog's visit. Or, better still, the dog should be put in kennels.

Many people, instead of putting their cat in a cattery during their holiday, hire a cat sitter or ask a neighbour to feed the cat. Most cats will tolerate this and are probably happier having the run of their normal territory. Occasionally, however, a nervous cat is very upset by the intrusion of a strange human into its home or is upset because the neighbour brings in the smell of her own cats. The resident feline makes its anxiety known by peeing in the house. If this is the case, a cattery will be the preferred arrangement in the future.

Other sources of stress for a sensitive cat are builders in the house (put your cat into a cattery), builders next door, or very noisy road works in the vicinity. The sound of dogs barking or the appearance of a new dog on the housing estate are other sources of stress.

CAT TIP

Before the homecoming of a new baby, it is a good idea to take home a handkerchief smelling of both baby and mother and place it in the cat's bed. This starts the mixing of scents that will reassure the cat that the baby is part of the family. Installing a Feliway® diffuser to diminish the cat's anxiety will also help.

Even a new baby may upset a resident cat. They will need time and understanding to adjust. There is a myth that cats suffocate babies by sleeping on their heads and many cats are dumped in rescue shelters as a result. There is no need. Sensible new parents will make sure that babies only get to meet pets – whether cats or dogs – under supervision.

Sensitive cats may take time to adjust to new flatmates or new partners. Partners who are cruel to cats will never be forgiven by the cat of the household. We cat lovers know the answer to this. Rehome the partner, of course!

CHAPTER ELEVEN

Running a Cat Care Home

'As I look back upon it, Calvin's life seems to me a fortunate one, for it was natural and unforced... The poet who wrote so prettily of him that his little life was rounded with a sleep understated his felicity; it was rounded with a good many. His conscience never seems to interfere with his slumbers.'

Charles Dudley Warner, 1829–1900

Cat, like humans, have benefited from better medical care and now they live longer and healthier lives. Thus many of us have to organize care not just for an elderly relative, but also for an elderly or disabled cat. Both elderly humans and elderly cats may be quite demanding and, at times, difficult – though in my opinion elderly cats in general complain less.

CAT FACT: Nowadays, thanks to better care, cats in their late teenage years are common and cats in the twenties are not as unusual as they used to be. In *The Guinness Book of Pet Records* the oldest cat was a female tabby, Ma, who lived to be 34 years and 5 months. When I asked newspaper readers the age of their old cats, the oldest lived to be 32. She was an ordinary moggy who had lived on a farm all her life. The Feline Advisory Bureau, a charity which has excellent information online at www.fabcats.org, suggests that a 10-year-old cat is the equivalent in age to a 56-year-old human, a 15-year-old cat is like a 76-year-old human and a 20-year-old cat is the same age as a 96-year-old human.

Probably the most obvious reason for cats living longer is down to vaccination, as a huge proportion of cats used to die of infectious diseases before the age of ten. Neutering is the other life-saver. Female cats burdened with three or four litters of kittens a year would also die early. Tabby, a little white-and-tabby cat whom I knew, had a litter of kittens followed by a second litter almost immediately because her owners were slow to get her spayed. She died shortly after the birth of the second litter of sheer exhaustion.

Entire tomcats used to become infected with serious diseases like FIV (feline immunodeficiency virus), which can be transmitted by biting during severe fights. Unspayed females suffered a similar fate, as the virus can be passed on when the tomcat bites her in the nape of her neck – part of the normal mating procedure. Neutering has not only reduced the number of unwanted cats, it has saved the lives of existing ones.

Cats, like humans, slow down in old age. I've heard it suggested that cats live longer than dogs because of this ability to do very little! They sleep a lot. One conscientious feline researcher did a day's study watch on an elderly cat, staying awake and noting all his movements for a full 24 hours. While she struggled to stay awake, the cat slept in peaceful serenity for 16 of those hours!

CAT TALE: Lucy, a beautiful tortoiseshell rescue cat, didn't have the best start in life but she has lived into her twenties – an immensely good age for a cat. She and her companion cat, Barney, were abandoned in a cattery when her owners went on holiday and never reclaimed their cats. Because the local rescue centre was full, the cattery owner, Gill, adopted the two cats.

Lucy was much more relaxed than her companion Barney, who seemed terrified of everything. It was just as well that she was laid back. 'Barney was forever jumping out on her from behind bushes or curtains,' says Gill. 'He was very much Top cat.'

When Barney died at the age of 17 from kidney problems, Lucy found her voice. 'We had never heard Lucy make a sound until we brought Barney's body back from the vets for burial in the garden. Lucy looked at him, licked him and started miaowing,' recalls Gill. 'She has never looked back and is incredibly vocal now.'

Lucy has been given excellent care throughout her life. She was never allowed out at night. 'We spend far longer choosing her food in the supermarket than we do our own,' admits Gill. 'The only operation she has ever had was to remove all but one of her teeth. This did not affect her appetite and she still eats well.'

At the age of 23 Lucy was still enjoying going out in the garden in fine weather, though she no longer goes very far. She has been fortunate in her devoted owner.

Happy the elderly cat that is well insured; happier still the owner...!

What do elderly cats need for a happy life? Just as we need a good doctor, they need a good vet. Owners who have prudently insured their cat with a good insurer, who promises lifetime care without increasing expense, will now be in the happy position of not having to worry about cost. Unfortunately, it will be too late to insure a cat once it achieves old age. In the UK, however, there are charities that will treat animals for free for people who are on benefits.

Vaccinations are important for a long life. Even for an indoor cat, vaccination for the most common infectious diseases is needed in case the cat is ever allowed out of the home. A vaccination schedule may not need to include *all* the diseases, nor need it necessarily be done yearly. A good vet will discuss this intelligently with the client.

Regular check-ups are as useful for elderly cats as they are for humans. There are some very common diseases in elderly felines such as hyperthyroidism (over-active thyroid) and kidney disease. These can be picked up early. Nowadays vets can take blood-pressure readings for cats, while in the past many older cats lost their sight because of high blood pressure.

CAT TIP

If an elderly cat starts making unusually loud yowling noises, take him to the vet for a check-up. Weight loss (despite a good appetite), restlessness and crying are symptoms of hyperthyroidism, a common disorder which can be easily diagnosed and treated.

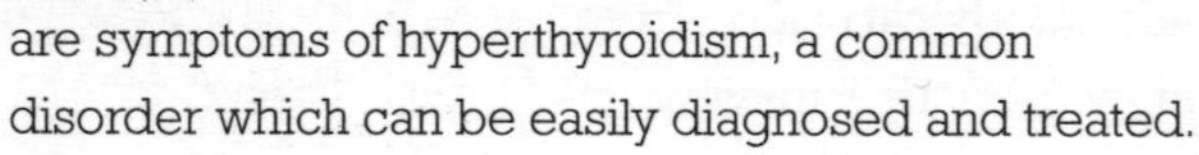

Stiff limbs and sometimes painful joints also afflict both the human and the feline after a certain age. An elderly cat that suddenly starts biting when being picked up may well experience pain when handled. Arthritic cats benefit from prescribed pain-killing drugs just like we do, though over-the-counter drugs like aspirin, which are completely safe for humans, can kill cats. There are also various joint supplements available, but it is very important not to use these without veterinary advice. It is possible to overdose a cat with cod liver oil, for example. Arthritic cats need help with grooming as they can no longer wash the difficult-to-reach areas.

Ramps of all kinds can help the elderly move upwards to the armchair or the bed. Heated beds or pads are a comfort to aching limbs. 'The whole house looks like a cat hospital!' said my exasperated husband in the days when elderly Fat Mog was still with us. Indeed, I have to confess to a ramp in the living room leading to the sofa, one in the bedroom for the bed, one for access to my desk, a heated tunnel (alas no longer made) in the bedroom, a heated bed on my desk, and a bed warmer for the living room! I stopped short of buying a heated waterbed, on the grounds that it cost about £500.

CAT TIP

If your elderly cat has become grumpy and lazy, it may be more than just old age. Cats with arthritis are often not visibly lame. Signs of arthritic pain are more time in the cat bed, hesitation in jumping on the bed or your lap, a matted coat because of less grooming, or a tendency to bite when handled. The vet can prescribe painkillers and you can buy heated beds that will stay warm throughout the night, which they find comforting.

All elderly cats should have a litter tray indoors. It is cruel to expect a geriatric cat to walk out into the cold wind and rain when he needs to relieve himself. In old age, knowing that they can no longer defend themselves, they may also be much more frightened of neighbouring cats. A litter tray indoors allows them the warmth and privacy they need.

CAT TIP

Make your own cat ramp using a strong piece of wooden plank which is smooth-edged so that splinters will not get into elderly paws. Then place strips of wood across it so that the cat can get some purchase with his feet – rather like the ramps that lead up to hen houses. Or simply cover it with fluffy bedding like Vetbed. The ramp will need a solid base at one end so that it does not slip on the floor.

Too fat or too thin? Weight control for the elderly

Overweight cats, just like overweight humans, are more vulnerable to arthritis and conditions like diabetes. Sometimes a programme of weight loss, under a vet's supervision with regular weighing, can have a remarkable effect on feline well-being. Cats should never be starved into weight loss, as this can be life-threatening. All cats require a very careful diet plan.

Weight loss is more common than weight gain in the elderly feline. Most old cats are thin and a bit scruffy. As the digestive system is probably less efficient in old age, there are now diets specially formulated for older cats on sale in the supermarkets. Vets can also supply special prescription diets of all kinds. It isn't sensible or even very kind to feed older cats too much human food; a good complete diet from a tin or a packet is actually better for them than a diet of cooked chicken or steak tartare, particularly in old age.

Teeth and hair – it's a downhill slope!

A surprising number of people don't realize that cats need dental care in old age just as much as humans do. The cat that goes towards her food bowl, takes a bite or two and then retreats shaking her head, paws at her mouth, or who has breath that smells, is a cat with dental problems. In theory we all should be cleaning our cats' teeth regularly. In practice, it is much more difficult to do this than it is to clean dogs' teeth. The teeth are further back in the head and even with a compliant cat that has had dental care since kittenhood, it's hard getting the brush so far back into the right area.

The alternative to teeth brushing is feeding something that will be crunchy enough to clean the teeth. Tinned or semi-moist food is useless for keeping teeth clean. Dry food is a little better but vets can prescribe dry food that comes in large pieces and a small proportion of this every day will help with teeth. I use it as training treats for my cat. There are also chews that can be helpful. Of course, crunching up the bones and sinews of mice or young rabbits is nature's way of cleaning the teeth, but elderly cats are less likely to be efficient hunters.

Good vets check the teeth when they are giving vaccinations. More complicated dental investigations have to be done under anaesthetic, though, and for persistent dental problems it is well to ask your vet to refer you on to a specialist dental vet. Teeth problems can be a sign of serious underlying disease and must always be investigated thoroughly. Gum disease is sometimes a legacy from earlier cat 'flu.

Cats with bad teeth find it difficult to groom themselves. Even without dental trouble, most elderly cats lose the sheen on their coat. Their hair becomes less thick and their coat can become matted. Even short-haired cats may have tiny mats on their back because they can no longer twist round to groom themselves there. It is important to groom elderly cats as mats can tighten and make the flesh below quite sore. I have found that they enjoy the use of a soft brush or the special rubber Zoom-groom when they will not tolerate tougher brushes.

Living with a disabled cat

When a human loses a leg, it's a tragedy. When cats lose a leg, they manage wonderfully well. The Queen of Sheba, a cat that lost her leg in a car accident, demonstrated that she could climb a high garden wall by legging it up a shrub nearby. Her owners, who were anxious to keep her away from the road that had cost her a leg, then removed all the shrubs near the wall. Sheba simply vaulted over. She would go to the centre of the garden and take a good run at the wall, With a huge leap, she would clear it – 'like pole-vaulting without the pole,' said her owners.

Cats may go blind in old age, sometimes because of untreated high blood pressure. They manage remarkably well, not least because their whiskers help them detect obstacles in their path. In this situation, owners should be careful not to move the furniture unnecessarily and should keep everything in its familiar place. It's also helpful to remember to speak to your cat before touching him – otherwise sleeping cats will wake up abruptly. Blind cats should be confined to the house, or at least to a well-fenced garden. Occasionally relationships with other cats break down – probably because the blind cat cannot read body language.

Because cats communicate so much by scent and body language, deaf cats seem to manage well with other cats. They too should live inside the house only, as their lack of hearing makes them vulnerable to traffic on the roads. If you decide to let them go into the garden, call them in with signals from a torch. They may also be startled when you come up behind them, as they cannot hear you coming. My own deaf cat, Fat Mog, became much more emotionally dependent upon me and developed a rather raucous miaow.

CAT TALE: Dinky, a ginger-and-white disabled cat, owes his life to his owner, Sam, two times over. As an orphaned kitten he was taken into the care of Cats Protection, the biggest cat charity in the UK. There he was bottle-fed by Sam and her mother until he grew up into a handsome adolescent.

But at the age of nine months he began to go wobbly, and then his back legs gave way completely, as did his bodily functions. Some owners would have put him down, but Sam didn't give up. She helped express his urine and faeces every day. Finally, the vet diagnosed a benign tumour on his spine and operated to remove it.

Dinky was nursed back to strength as well as health by being taken to hydrotherapy. In the days before the operation, when his back half was paralysed, his back legs had lost muscle mass and were severely wasted. Even though the operation was a complete success, something had to be done to help him get his strength back.

'My dog went to swimming therapy, so we knew it worked,' said Sam, who is an animal care assistant at a veterinary surgery. 'We decided it could work for Dinky too so we bought a life jacket (for a dog) for him at a garden centre.' Twelve weeks after surgery he started in the pool.

To begin with, Sam and her parents simply got Dinky used to being in the hydrotherapy room. They took him in his carrier and he stayed there while their dog swam. The aim was to get him used to the new surroundings before he went into the pool itself.

At first the hydrotherapist simply placed him in the heated pool and, because he was wearing his life jacket, he floated without any difficulties. Slowly he learned to swim – and learned to enjoy it too. He would relax and float with his head on the side of the pool, as well as swimming.

'From being skin and bone, weighing only 1.8 kg (3.9 pounds), he went to 3.4 kg (7.5 pounds),' said Sam. 'He really enjoyed it and now when he is at home he will jump into the shower or the bath.'

Nine months later, Dinky's future seems assured. Once he was a cat unable even to walk; now, thanks to Sam's devoted care, he is fully functional – and can even mouse.

Can cats suffer from senile dementia?

Elderly cats sometimes show signs of confusion. They may sit in a corner as if they do not know where to go, forget to eat, fail to use the litter tray, make yowling noises at odd times, or sit persistently miaowing near an already-open door. Do not assume your cat is going barmy, he may just be suffering from a disease. Ill cats do things which are out of character, just as ill humans do.

Proper diagnosis and treatment for what is wrong will usually restore the cat to his proper mental health. However, if after tests nothing seems wrong, it could be that the cat has dementia and not enough oxygen is reaching the brain. We now also know that cats can develop a disease like Alzheimers in humans, and their behaviour can be similarly odd.

A better diet, rich in anti-oxidants and designed for elderly felines, can help. There are also drugs, mostly used for dogs, which may be suitable for cats. Cats that seem to get lost should be kept inside the house. Elderly cats that are losing it also become very anxious, but they can benefit from a Feliway® diffuser to help promote a calmer frame of mind. Litter-tray use can sometimes be reinstated by putting down more trays or making sure that the tray has low sides which are easier for an elderly cat.

CAT FACT: More than a quarter of pet cats develop some kind of illness or disorder between the age of 10 and 14 years, and after the age of 15, one in two cats will have something wrong with them. After the age of 20, most of them will show a little bit of memory loss and will be less active and supple. Just like us!

CAT TALE: George is a handsome black-and-white cat nearing the age of 20. He has the thin and rather scruffy look of an old cat, but his eyes are bright and he can see perfectly well even though he is a little deaf. In his early years he had spent time out of doors mousing and checking his territory. Now, in his last few years, he spends a lot of time sleeping on the front windowsill, where the sun slants in during the day.

He has two beds there – one with a heater and one without. If he gets too hot, he walks over, past his favourite teddy bear, to the cooler bed without a heater; then, if he gets too cold, he walks back again to the heated bed. To help him get to his windowsill, his owner, Joode, places a cushioned stool near the sofa. George's route takes him from the stool to the sofa, up the arm of the sofa, onto the back and from there to the windowsill in a series of small steps.

Just before his twentieth birthday, George had a day when he kept falling over. Every now and again his legs would go.

Joode thought it might be time to say goodbye to him, but after a couple of these attacks George recovered well enough to relieve himself on a newly planted flower border. The next-door doctor said it must have been transient ischaemic episodes – that is, mini-strokes that occur without lasting brain damage. George has been with the same family for all his life – though for four of his years he was with Joode's daughter while she was in the USA. His is a life well lived.

When it is time to say goodbye

Letting go is always painful, but the last gift we can give a cat is the blessing of an easy death. Here are some questions to ask if you think your cat might be nearing the end of his life:

- Is my cat still enjoying his food?
- Does my cat still play, if I invite him to do so?
- Is he able to control his bowel or bladder?
- Can he get to his litter tray, to his food bowl and to his bed?
- Can he still wash the easy-to-reach areas of his body?
- Is he suffering from frequent fits, or vomiting, or other disorders?
- Is he in pain?
- Does he require repeated veterinary interventions or care procedures such as injections?
- What does my vet recommend?

Do not strive unnecessarily to keep alive a very old or an ill cat. All cats hate going to see a vet and too much veterinary treatment can be as cruel as too little. There must always be an assessment of whether a cat's quality of life will suffer. Vets will give their advice and an owner should always take this very seriously. Do not be selfish: put your cat's welfare before your own feelings of loss. This is the last gift you can give your cat.

CONCLUSION

So Why Do We Still Love Them?

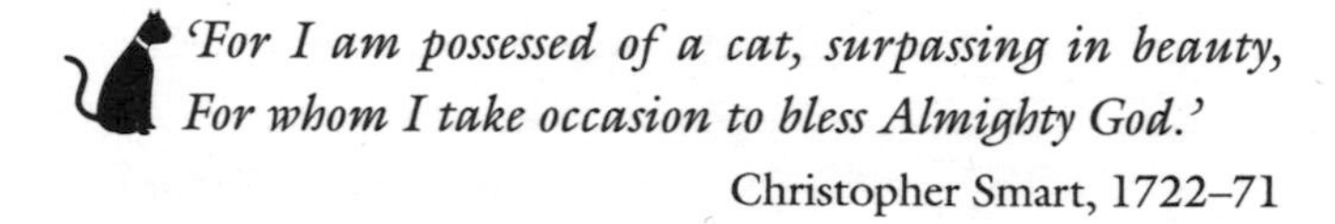

'For I am possessed of a cat, surpassing in beauty,
For whom I take occasion to bless Almighty God.'

Christopher Smart, 1722–71

The love between humans and cats is, like any love affair, an exciting and intriguing mystery. Relationships are meant to be best when the couple have a lot in common. Humans and cats don't share this; there is a huge gulf between the habits, priorities and activities of their species and ours. Yet this surprising relationship works well most of the time. We both share feelings of sadness, fear, anger, playfulness, happiness and ecstasy, but our thoughts are very different. Nor can we ever ask our cats what is going on in their heads. Their inner musings remain impenetrable.

So how have cats become our best friends? (In many countries they even outnumber dogs.) They don't fetch and carry for us. They don't (usually) obey us. They go off out of the house on their own hunting and exploring expeditions and we have no idea where or what they are doing. Some cats treat their homes merely as bed-and-breakfast establishments, coming in for meals and sleep and taking very little notice of their humans. Others, like Roofer in Chapter 5 (see pp. 97–9, 102), are often deeply unpleasant to their human companions, scratching or biting at the slightest excuse.

Yet they bring to our lives beauty and gracefulness. There is no dog that can compare in beauty and elegance to a cat. From the silence of their soft padded paws to the low rumble of their purr, they are a miracle of grace. They also bring into our homes all the wildness of Nature herself. Dogs have adapted themselves to human society; cats have remained themselves – small hunters equally at home by the fireside or out at night going about their mysterious ways in the moonlight.

No wonder we love them. However bad their behaviour (in our eyes), we still love them. Indeed, some of us love them to desperation point! Some of us, who have Cats-That-Do-Not-Do-Cuddles, yearn for them to give us more affection. Over and over again we make advances to them, advances that are always repulsed, with or without claws. The sight of our much-loved feline just stalking away in disgust after repelling our advances can be very upsetting. Yet we persist in loving.

I think the secret of cats is that they are like the romantic figure of Mr Darcy in *Pride and Prejudice*. We cannot, without deep cunning, manipulate or charm them against their will. The occasional glance of contempt from a cat only lands us deeper in love. We can't capture their hearts. Unlike dogs, even at their most giving moments, they hold back some inner part of themselves.

People who see animals as creatures to be used will never have much time for pet cats, though they may encourage the settling of feral cats in their stables or farmyards as mousers. Likewise, those who want a companion that can be ordered about will always prefer dogs. Dogs are jolly companions, anxious to please and even coming back for more after being abused by punishment. Dogs really do lick the hand that hits them.

Cats are less stupid. They are far too independent to work for us, except by mousing. They have learned, from centuries of ill treatment and cruelty, never to trust a human that offers punishment or abuse. They can hold a lifelong and self-protective grudge against those that frighten them. If home stops feeling safe, cats just leave home. They don't go back for more punishment. They are free spirits.

So when cats behave in ways that are difficult for us, we are the ones who have to change – whether it is stopping giving them affection they don't want, or putting down three extra litter trays. We can't change them, except by changing ourselves.

You never really *own* a cat. And that is the magic of cats.

USEFUL INFORMATION

My own website, www.celiahaddon.com, has information which is not in this book and is worth viewing. If I am not too busy, I respond to queries via the website.

Cat Behaviour Counsellors

For cat behaviour counsellors in the UK, go to the Centre of Applied Pet Ethology (www.coape.co.uk) or the Association of Pet Behaviour Counsellors (www.apbc.org.uk). Make sure you get a counsellor who is really experienced with cats.

Cat Deterrents

Spiked tree collars, called cat deterrent spikes, to stop kittens or cats climbing trees are available from Jacobi Jayne (www.livingwithbirds.com).

Cat Rescue

Cats Protection (www.cats.org.uk), the biggest UK rescue organisation for cats, also has some very useful information on topics such as feeding stray cats and cats and babies, for example. It is a good place to find out more about adopting a kitten or cat.

General Information

The Feline Advisory Bureau (www.fabcats.org) is a wonderful resource for information about all kinds of cat diseases. It also has useful information on inherited diseases in pedigree cats, cat breeding, fencing your garden, catteries, feral cat rescue, and much more.

Insurance

In the UK, Petplan's lifetime policy insures cats for life and gives benefits that may cover consulting a cat behaviour counsellor. Visit www.petplan.co.uk for more information.

Microchips

Petlog (www.petlog.org.uk) is Britain's largest microchip organisation and also provides a lost cat service.

Microchip Cat Flaps

PetPortes, the cat flaps that are programmed to respond to your cat's microchip, are available at www.petporte.com. These will stop neighbouring cats coming into your cat's indoor territory.

Nest Box Cameras

CJ WildBird Foods (www.birdfood.co.uk) sell bird nest boxes with camera attachments that play on your TV.

RSPCA

The RSPCA (www.rspca.org.uk) is the place to start if you know of people who are treating cats cruelly or if you come across an animal hoarder. You can report this anonymously. The RSPCA also has rehoming branches across the UK.

Synthetic Pheromone Sprays

Sprays, such as Feliway®, are now available from cheap and respectable online pharmacies.

Window Screens

Cataire (www.cataire.co.uk) provides a variety of window screens to help keep your cat safe and secure.

ACKNOWLEDGEMENTS

My thanks to Jim Vaissie of Howletts Wild Animal Park in Kent (http://www.totallywild.net/howletts/) for his help on the African wild cats there. More research into the behaviour of our cat's ancestor, the African wild cat, is still badly needed. Thank you too to Kathie Gregory (cat behaviour counsellor with the Centre of Applied Pet Ethology, and pedigree breeder of silver-spotted British Shorthairs), whose kitten expertise much exceeds mine. A special thanks to Professor Peter Neville, whose influence has led me back to serious study! In pursuit of this, I have had help from Roger Coftin and Rachel Leather. Finally a thank you to tabby-and-white William, the late playful black George, little fat black Mog, and portly black-and-white Ada. If she hadn't turned up in my back garden with her kitten, this would probably be a book about gardens, not cats.

INDEX

Index